AMPLIFIED LEADERSHIP

PARTNERING WITH AI TO DEVELOP PEOPLE, PURPOSE, AND PERFORMANCE

RACHEL E. WALTER

ISBN 979-8-9946269-0-0 (paperback)

ISBN 979-8-9946269-1-7 (hardback)

ISBN 979-8-9946269-2-4 (Kindle)

❀ Formatted with Vellum

Thank you to every person who has participated in my journey in any way and in the lessons that I have learned. Many, although not all, are mentioned in this book. Whether your name occurs within these pages or not, please know that you have helped me.

TABLE OF CONTENTS

INTRODUCTION

Too many leaders become mired in short-term thinking, focusing on quarterly earnings or immediate wins. However, the true measure of leadership lies in the enduring impact one leaves behind. Did you cultivate stronger leaders? Did you instill a culture that transcends your tenure? Did you position the company for sustained success?

Legacy is fundamentally pragmatic. Leaders who effectively multiply their positive influence through talent development, cultural stewardship, and tangible results extend their impact far beyond their physical presence within the organization. Every individual leaves an indelible mark on the people they encounter. These marks can either foster growth and development or, conversely, erode and diminish potential. By prioritizing the creation of a positive legacy, leaders actively shape the future. While the principles of legacy leadership are universally applicable, this book is specifically tailored to the vantage point of enterprise-enabling leaders – those who may not be directly responsible for sales quotas but bear the responsibility for enterprise-wide outcomes.

My career path did not begin with the intention of authoring a book on intentional leadership development through specific activities. Instead, it commenced with an entrepreneurial venture rather than a traditional corporate trajectory. My co-founder, Laura Klopp, and I established SpiraTech, a graphic design and technical writing firm. We diligently pursued clients, crafted documentation, and designed graphics late into the night. I quickly learned that entrepreneurship provides invaluable lessons in business survival that no classroom can replicate. Cash flow was not merely a theoretical concept; it represented the stark difference between meeting payroll obligations through loans or relying on company reserves.

Subsequently, I transitioned to Mercury Mercruiser as an engineering documentation manager, leading a team of technical writers. This role required a shift in perspective, from simply executing tasks to managing personnel and implementing strategies not of my own creation. I had to think differently. Writing manuals was not enough. I had to innovate to better support customer needs, to align a diverse team, to manage deadlines, and to deliver what engineers, customers, and executives needed.

Later, at Fire Protection Publications / IFSTA, I transitioned into program management for fire and emergency services training. This role served as a crucial bridge between technical expertise and client requirements. The programs we designed carried significant weight, as firefighters relied on clear, reliable training to protect themselves and others. I honed my program management skills and developed a strong sense of accountability, recognizing that the outcomes of our projects extended beyond mere business metrics.

My career then led me to Hilti, a global construction tool and services company. I initially focused on sales training and eventually assumed responsibility for learning and development across the entire business. Here, my experiences coalesced, and I witnessed firsthand the pervasive myth addressed in Chapter 1: that learning is often undervalued and excluded from strategic decision-making. Training was frequently dismissed as a cost center, and I was determined to challenge this perception. By reframing learning as a catalyst for sales conversion, leadership readiness, and accelerated change adoption, we transformed the organization's perspective.

Executives began to recognize L&D not as a mere service function but as a key driver of growth.

Today, I serve as Chief Innovation Officer at **ansr**source. My career trajectory has encompassed a broad range of experiences, allowing me to develop both deep expertise and a comprehensive understanding of various business functions. I oversee innovation across products, services, and strategy, collaborating with sales, product development, operations, marketing, and finance teams. In this role, I observe recurring patterns similar to those I encountered in L&D: a lack of emphasis on developing future leaders and a tendency among functional leaders to focus on activity metrics rather than maximizing their impact on the business.

This professional journey has profoundly shaped this book. I have experienced both sides of the table, as both an individual striving to gain influence and an executive making investment decisions. I have witnessed firsthand what strategies succeed, what approaches fail, and what practices create lasting value.

The market is saturated with generic leadership advice. This book does not offer superficial recommendations such as "communicate better," "develop your people," or "cultivate executive presence." Instead, it provides leaders with the clarity they need to connect their daily activities to tangible, impactful results. Too often, leaders are marginalized because they communicate using functional jargon rather than business-oriented language, inadvertently fostering a narrow, siloed perspective among their teams.

- Sales leaders focus on call volume instead of closed deals and the factors that drive successful conversions.
- Marketing leaders emphasize impressions rather than pipeline conversion rates.
- Product leaders prioritize features over user adoption.

L&D leaders often discuss training hours instead of revenue acceleration, while operations leaders focus on cost reduction rather than ensuring future-oriented, sustainable profitability. These are activity metrics, easily tracked and reported, but they do not establish credibility at the executive level.

Executives, whether in the boardroom or on an earnings call, communicate in terms of growth, cost, risk, and value. If you cannot connect your work to at least one of these outcomes, your influence will be limited.

WHO THIS BOOK IS FOR

This book is based on my experience as a Chief Learning Officer and Chief Innovation Officer. It is written for leaders of enterprise-enabling functions whose work drives growth, performance, and culture across an organization, even without direct revenue responsibility.

Specifically, this book is for:

- CLOs and CHROs seeking to demonstrate how talent and culture fuel strategy.
- CFOs and COOs aiming to tie operational excellence directly to profitability and resilience.
- CMOs, CTrOs, and CInOs who want innovation, transformation, and brand growth grounded in measurable outcomes.
- CEOs and General Managers who want every function to communicate in the language of enterprise results.

If you have ever left a leadership meeting frustrated that your team's work did not resonate or wondered why budgets were cut while other areas thrived, this book is for you.

If you aspire to have your function viewed not as a support service but as a strategic enabler, and if you want the impact of your leadership to extend beyond your tenure, this book is for you.

HOW THIS BOOK WORKS

Each chapter builds upon the previous one.

- Chapter 1 dismantles the myths that sideline leaders.
- Chapter 2 transforms activities into outcomes.
- Chapter 3 sharpens leadership business acumen.
- Chapter 4 elevates talent development as a responsibility of every leader.
- Chapter 5 translates activities into the language of results.
- Chapter 6 provides the tools for leaders to forge powerful alliances.
- Chapter 7 embeds culture into leadership practice.
- Chapter 8 extends leaders' reach with technology.
- Chapter 9 anchors leadership through governance.
- Chapter 10 sustains your leadership influence and legacy.
- Chapter 11 translates the techniques into enterprise leadership.
- Chapter 12 wraps it up with a look at your leadership legacy.
- Appendix: Self-Assessment equips you to focus your growth.

Read the book sequentially to understand how the chapters form a system of leadership that earns influence and drives results.

Understanding that leaders often have limited time, a brief Self-Assessment is included to help you identify your strengths and areas needing immediate attention. If governance is your biggest gap, proceed directly to Chapter 9; if culture is the challenge, go to Chapter 7. This book is both a path and a toolbox. Follow it step by step or use the tools you need most urgently.

This book is grounded in real-world experience, not abstract theory. Each chapter draws upon my own professional journey, as well as the experiences of clients and colleagues. You will encounter stories of leaders who faltered due to a focus on irrelevant metrics and others who flourished by aligning their work with tangible business outcomes. You will also see how organizations across diverse industries have leveraged these principles to accelerate growth, reduce costs, mitigate risks, and enhance value.

Furthermore, each chapter concludes with AI prompts designed to stimulate critical thinking. As a Chief Innovation Officer, I utilize AI daily and want to share my methodology. AI can sharpen your insights, identify patterns, and reveal overlooked questions. In this book, AI serves as your thought partner, empowering you to test the concepts presented against your own company's data, language, and strategic objectives. The leaders of tomorrow will possess both a deep understanding of their business and the ability to harness AI for accelerated and more intelligent decision-making.

MY COMMITMENT TO YOU

I offer no platitudes, nor do I suggest that leadership is easy. Leadership is a challenging endeavor, often imbued with complexity, politics, and a lack of immediate recognition. Yet, it is also profoundly rewarding, for leadership transcends individual ambition. It is about the people you serve, the success of the business, and the lasting impact you create.

My motivation for writing this book stems from my own experiences; both the frustration of being marginalized and the satisfaction of driving transformational change. I have navigated the challenges of entrepreneurship, managerial responsibilities, program leadership, learning and development, and driving innovation.

Through these experiences, I have learned that leaders who connect their work to business outcomes, invest in the development of others, communicate in the language of results, and build enduring systems are the leaders who leave a lasting legacy.

This book is designed to help you become that leader.

1

THE SEAT AT THE TABLE MYTH

Throughout my career, I have observed many talented heads of functional departments succumb to a common pitfall: seeking influence for its own sake. They aspire to a seat at the table, but when given the opportunity to contribute, they speak in activities and acronyms for their departments.

Meanwhile, their peers, the ones with influence, are focused on growth, margins, and market share. Consequently, the functional head often receives polite acknowledgment, after which the discussion reverts to core business matters. For revenue-generating leaders, impact is more readily apparent. For enabling leaders, impact requires translation. Earning a seat at the table is less about authority and more about effective communication.

The prevailing myth: **CLOs lack influence because CEOs undervalue learning.**

The reality: **Many functional leaders fail to position themselves as drivers of business outcomes.**

This challenge is certainly not unique to CLOs; sales leaders, logistics leaders, CTOs, CROs, and others often make a similar error. They concentrate on their specific area rather than the broader organizational objectives, focusing on departmental activities without connecting them to overarching business goals. Companies thrive when leaders collaborate across functional boundaries.

WHAT IT TAKES TO EARN A SEAT AT THE EXECUTIVE TABLE

Chief Learning Officers don't lack ambition. Virtually every functional leader aspires to be recognized as a strategic contributor, and CLOs are no exception. They seek to influence the company's strategic direction beyond the confines of classroom management or compliance training. However, when presented with the opportunity to address the executive team, many CLOs revert to discussing learning hours, completion rates, or employee engagement scores, quantifying the number of individuals processed through training programs or the staffing levels required to reach every employee.

These are not business outcomes; they are activity metrics.

I have observed training managers presenting elaborate dashboards replete with employee training percentages, course launch statistics, and learner satisfaction scores. While their efforts to gather this data are commendable, the CFO remains unconcerned with the number of employees who clicked through a module. The CEO does not base strategic decisions on post-training "smile sheets," and the COO will not authorize investment based solely on positive learner feedback.

These leaders prioritize revenue growth, operating margin, customer retention, and speed to market. If L&D cannot articulate its impact in these terms, its strategic value remains invisible.

This challenge extends beyond the L&D function.

Sales leaders sometimes flood dashboards with activity-based metrics such as calls made, demos booked, and emails sent. While these figures may convey a sense of activity, the executive team requires corresponding improvements in win rates, customer lifetime value, or strategic account expansion to validate their effectiveness. A robust pipeline of conversations does not necessarily translate into a robust pipeline of closed deals.

Product leaders often showcase the number of features delivered, bugs fixed, or user stories completed within a sprint. Executives, however, fund development based on anticipated outcomes, seeking tangible results such as customer adoption, retention, and competitive differentiation. A comprehensive release log does not guarantee enhanced customer loyalty or the generation of new revenue streams.

Across these functions, a common pitfall emerges: leaders prioritize reporting easily measurable metrics over those that genuinely impact the business. They present activity metrics in discussions where outcome metrics are expected. At best, these activity measures serve as leading indicators of future business results; at worst, they are a distraction. These parallels highlight a pervasive issue extending beyond learning; across the C-suite and beyond, leaders often fall into the trap of reporting what is easy to measure instead of what drives business outcomes.

My entry into the CLO role differed from that of many peers. My career trajectory, shaped by education and diverse roles, began firmly rooted in business. I have managed P&Ls and navigated cost center allocations and overseen budgets where profit margins were paramount. I pursued an MBA after my entrepreneurial endeavors, driven by a need to understand the mechanics of organizational growth, survival, and success. Consequently, I have always viewed learning as intrinsically linked to business results. However, I quickly observed that many leaders in the corporate world, particularly within HR or L&D, did not share this perspective. While often well-intentioned and focused on people, they overlooked a fundamental principle:

Employees retain their positions only when the company performs well enough to sustain their employment.

While this may seem blunt, it is a fundamental truth. Organizations do not exist solely to provide learning opportunities, CRM dashboards, or new product features. Their primary purpose is to deliver value to customers, generate profit, and achieve sustainable growth. When the business thrives, employee development can flourish. Conversely, when the business falters, budgets are invariably cut across operations, learning, sales enablement, and product innovation.

However, leaders who exclusively emphasize functional outcomes such as engagement rates, call volumes, or features shipped inadvertently reinforce the perception that their function is merely a "nice-to-have." They miss the opportunity to demonstrate how these outcomes contribute to the company's financial success.

These leaders often communicate ineffectively, framing their contributions in terms of inputs rather than outcomes, focusing on activity rather than impact. They maintain fluency in their function-specific metrics but lack proficiency in the broader vocabulary of business. Engage in a conversation with a functional leader from another area and you will likely understand only a fraction of what they convey as they will employ function-specific jargon and discuss topics that, while perfectly clear to them, lack relevance to your context.

Until this dynamic shifts, these leaders will continue to operate on the periphery, failing to connect their work to the business issues that stakeholders value most. Unfortunately, this also means that their teams will perceive them as modeling a role that operates on the sidelines.

The Metrics That Missed the Mark

In France, I collaborated with a training manager, Elodie Bernard. Due to government reimbursement policies for training hours, Elodie meticulously tracked all training delivery. Her Training Operations Manager, Louis Gaston, compiled this data into a monthly dashboard, believing he was providing her with the necessary tools to demonstrate control and rigor.

However, Louis was unaware that Elodie only required this data annually for government filings and did not utilize the monthly reports. To avoid discouraging him, she allowed him to continue their production.

This had an unintended consequence: these reports, filled with charts on training hours, were included in the deck reviewed monthly by the CHRO and CEO. The implicit message conveyed was that training was being managed as a compliance function, emphasizing hours tracked rather than business outcomes achieved.

Consequently, the CEO perceived that Elodie did not fully grasp the critical importance of customer retention and maintaining a competitive edge through technology. Viewing training as disconnected from the business, he bypassed her, occasionally collaborating with the HR Head to engage external providers for strategically important programs, without consulting Elodie or her team.

Elodie's performance was not due to incompetence, but rather a response to the system's emphasis on tracking hours. Because her function communicated in training-centric terms rather than business-oriented language, the CEO did not perceive her as a strategic partner, resulting in the marginalization of training.

This same pitfall extends to other functions, even those traditionally considered core to the business.

A Chief Marketing Officer (CMO) may present marketing dashboards showcasing thousands of leads generated and millions of impressions captured. However, when the CFO inquires about the number of leads that converted into qualified pipeline opportunities, the response is vague. The CMO may explain that this data falls under sales and is inaccessible. Executives invest in growth, not merely awareness. Without linking activity to conversion, marketing appears busy but lacks strategic value.

A Chief Operations Officer (COO) might highlight utilization rates and the percentage of hours billed, without connecting these metrics to profitability or long-term client loyalty. In one instance, a COO heavily relied on contract labor to maintain high utilization rates, which appeared efficient on paper. However, customers experienced churn, innovation slowed, and new business referrals declined. Executives need to understand how operations will adapt to be smarter, more effective, and more innovative, enabling the company to command premium rates and garner client recommendations.

These scenarios may appear distinct, but they share a common root cause: reporting on activity instead of outcomes.

During an annual budget cycle, I recall a Chief Learning Officer (CLO) at a mid-sized technology firm who requested additional funding. She presented a detailed presentation highlighting the number of courses launched, employees trained, and positive survey results from participants.

The CFO listened attentively before posing a critical question: "If I approve this budget, how will it contribute to acquiring more customers or reducing costs in the coming year?"

Unfortunately, she could not provide a direct answer. Her request was solely tied to training volume because she saw her team overworked trying to deliver training programs. She lacked a clear connection to revenue, retention, or risk mitigation. Consequently, the budget increase was denied, and more significantly, the perception spread among the executive team that learning could not justify its investment in tangible business terms.

Learning leaders must dispel the misconception that CEOs undervalue learning. Continuing to present training metrics, engagement rates, or completion rates in executive meetings without demonstrating a clear return on investment will only result in continued budget cuts during times of business pressure.

Executives are not short-sighted; they understand that during challenging periods, a business's primary focus is survival, which hinges on revenue, margin, cash flow, and growth. Any function that cannot directly contribute to these outcomes is quickly deprioritized.

This principle extends beyond Learning and Development (L&D):

- Marketing leaders who report lead volume without demonstrating lead conversion to opportunities risk budget cuts during periods of slow growth. Executives prioritize revenue generation; marketing leaders should highlight conversion metrics.

- Operations leaders who emphasize utilization rates without linking them to innovation, profitability, or client loyalty risk losing credibility. During times of pressure, efficiency metrics without demonstrated effectiveness or future orientation appear inconsequential.
- Sales leaders who track activity metrics without focusing on strategic account expansion risk being relegated to order-taking roles rather than shaping growth strategy.
- Product leaders who highlight release counts without demonstrating adoption or retention may be perceived as R&D overhead rather than drivers of competitive advantage.

When leaders primarily communicate within the confines of their function rather than using the language of business, the following consequences arise:

1. **Budget Reduction:** Failure to demonstrate a direct contribution to business outcomes positions a function as a cost center, making it vulnerable to budget cuts.
2. **Erosion of Influence:** While still invited to the table, the leader's voice loses weight, and they are often relegated to providing updates after critical decisions have been made.
3. **Talent Impairment:** When functions are marginalized, organizations risk developing leaders who are unprepared for growth, employees who are disengaged from the mission, and teams that are unable to execute the strategy effectively.

When leaders shift their focus from functional outputs to connecting their work to tangible business performance, new opportunities emerge. The nature of the conversation evolves, transitioning from requests to "deliver a report" or "run a campaign" to inquiries about how the team can contribute to achieving the company's growth strategy. Instead of budget cuts, executives explore the potential impact of additional investment.

I have witnessed this shift firsthand. When Chief Learning Officers demonstrate business acumen, they transition from being perceived as service providers to being valued as strategic partners. When Chief Marketing Officers tie campaigns to pipeline conversion, they shift from defending their expenditures to driving the growth agenda. When Chief Operating Officers connect efficiency to innovation and customer loyalty, they transition from being measured solely on cost to being trusted to shape the company's competitive edge.

This transformation extends beyond job security or budget considerations; it speaks to purpose. Most leaders enter their fields with a desire to create value for customers, teams, and the business. That purpose is only realized when the company thrives.

Learning has the power to accelerate business results. Marketing has the power to position the brand for sustainable growth. Operations has the power to increase operational efficiency while simultaneously strengthening capabilities. Sales has the power to translate vision into tangible customer contracts. Product has the power to unlock new markets. Human Resources has the power to anticipate and adapt to market or industry shifts. And all functions accelerate results when they clearly and consistently connect their efforts to overarching business objectives.

The Research Behind the Myth

Data reveals a systemic pattern: many functional leaders are excluded from strategic conversations because they infrequently connect their work to tangible business outcomes.

Specifically, RedThread Research recently reported a 50% decline in L&D's involvement in business-strategy discussions over the past two years. This suggests that the fewer connections learning leaders make to genuine business impact, the less likely C-suite members are to include them in strategic decision-making processes. Consequently, L&D's contributions are often categorized with operational

HR functions like compliance, benefits, and payroll, rather than being recognized as a driver of growth, retention, or innovation.

Josh Bersin's research elucidates the underlying reason: fewer than 10% of organizations systematically link HR and learning data to key business metrics. If L&D cannot demonstrate how its work influences revenue, customer outcomes, or cost control, executives will likely assume it does not. And when that assumption solidifies, learning becomes a budget line item to minimize, rather than a strategic lever to accelerate organizational performance.

This often results in CLOs reporting to HR rather than directly to the CEO. When learning is positioned as a sub-function, it can become overshadowed by the more immediate operational demands of HR, such as talent acquisition, compensation, and compliance. While HR operations are undoubtedly critical, this can cause the strategic view of learning as a driver of business success to be lost. Andrew Hunt, Head of L&D in one of Hilti's Asian regions, exemplifies the strategic positioning of L&D, reporting directly to top leadership rather than within other functions. Upon rejoining Hilti, after leading several successful organizations, he brought a refined ability to weave compelling business narratives through the lens of L&D. This manifested in his success in developing training managers into key leadership roles by teaching them to communicate in the language of executives, crafting compelling narratives that connected directly to organizational goals. As a result, they became capable and confident unit leaders, driving organizational success. Andrew is building his legacy by developing business-savvy leaders.

The true measure of L&D's value lies not in training hours or engagement scores, but in the CLO's direct access to the CEO and their ability to articulate the company's strategy, risks, and opportunities. Just as a CMO reporting solely on click-through rates without linking them to qualified pipeline may miss the mark, finance executives demand concrete evidence of impact. For example, Nielsen's 2024 and 2025 Annual Marketing Reports and Insights publications underscore the marketing metrics that resonate most with CFOs and drive budget support. Researchers found that while impressions and engagement garner attention, CFOs prioritize metrics such as cost per lead and marketing-attributed revenue. These outcome-focused indicators directly link marketing investment to financial performance, enabling leaders to move beyond activity reporting and embrace business-aligned storytelling.

> Earning a seat at the table is not a courtesy; it is a testament to the recognition of your function as a key contributor to business outcomes.

The Myth Does Not Define You

By shifting the way you and your team engage with the business, you can unlock transformative change.

Business acumen is a force multiplier for any team. Whether you lead learning, HR, IT, analytics, or purchasing, or manage product development or sales teams, leaders who understand the intricacies of the business gain influence. They are trusted early in conversations and included in the decisions that shape the future.

Developing up-and-coming talent is the responsibility of every department. Talent development is not the sole domain of HR or L&D. Every leader must identify potential, stretch capabilities, and cultivate the next generation of leaders.

Every function can impact business results. My personal experience has shown that any function can (and should) serve as a lever for growth, culture, and performance.

This book is about empowering you to become the kind of leader who consistently delivers value and earns their seat at the table. Here's how to achieve transformational leadership:

1. Cultivate a comprehensive understanding of the business, transcending the limitations of your functional perspective.
2. Develop talent strategically to directly support and accelerate organizational objectives.

3. Communicate results effectively, emphasizing their impact on key business priorities.
4. Establish a strong network of allies to broaden your influence and create opportunities.
5. Shape the organizational culture to align daily behaviors with the company's strategic vision.
6. Leverage technology, particularly AI, as a force multiplier to enhance your overall effectiveness.

Earning a seat at the table requires demonstrating leadership and driving tangible business results. This book will show you how to do that yourself and how to leave a legacy of leaders who do that as well.

∾

LEVERAGING AI AS A STRATEGIC THOUGHT PARTNER

In this chapter, AI is not used to optimize output, but to help you examine how your function creates value in terms executives recognize and reward. Consider reframing your function through the lens of AI.

Prompt: "List five ways the [insert your function here] in [insert company name] impacts shareholder value." If this is your first use of this AI tool, be prepared adapt the prompt by first entering: "Ask me questions to understand my organization enough to be able to list five ways the [insert your function here] in [insert company name] impacts shareholder value."

For example, when applied to Learning & Development (L&D), the results may include:

1. Reduced employee turnover, leading to savings in rehiring and retraining expenses.
2. Accelerated onboarding, shortening time-to-productivity and boosting revenue generation.

3. Stronger leadership pipelines, minimizing the cost and disruption of external recruitment.
4. Targeted compliance training, mitigating the risk of fines and legal liabilities.
5. Upskilled teams, enabling faster adaptation to technological advancements and maintaining a competitive edge.

Reflection

- Which are currently evident within your organization?
- How can you strengthen these with data and evidence?
- Which would be most impactful to your CEO?

Incorporate one of these points into your next executive discussion. By linking your function to shareholder value, you shift the conversation from activities to strategic business outcomes.

In the chapters that follow, this same approach will be applied to business acumen, leadership pipeline, results communication, alliances, culture, and technology.

2

TRANSITIONING FROM FUNCTIONAL HEAD TO TRANSFORMATIONAL LEADER

I n the previous chapter, we addressed the misconception that functional leaders lack influence or that their voices are under-valued by CEOs. The reality is that many functional leaders focus on activities rather than transformative results. While metrics reflect our priorities, they can also create limitations. This chapter is about translation: converting functional requests into enterprise outcomes. Chapter 3 will strengthen the business acumen that makes those outcomes credible, and Chapter 5 will address how to communicate results with rigor.

Remaining within a functional perspective emphasizes activity. However, adopting an enterprise-wide view reveals the broader impact. For many enabling leaders, the key is not to create new KPIs, but to reframe existing ones in the language of the business. This chapter is about moving from delivery to transformation. Delivery focuses on fulfilling requests and reporting activity. Transformation focuses on changing outcomes that matter to the business.

In L&D, training activities include the number and type of courses offered, employee certification rates, and content delivery hours. Transformation, on the other hand, focuses on outcomes such as leadership readiness for expanded roles, the speed of new talent integration, and the capability of teams to execute strategy faster than competitors. By prioritizing transformation, you evolve from a service provider to a strategic business partner.

I will always remember when a key stakeholder invited me to a meeting regarding sales training. The department heads were discussing their training needs with my team, expressing a desire for a standardized program with modules on the sales cycle, presentation skills, and objection handling. It was the kind of generic sales training common in many organizations.

When I had the opportunity to speak, I posed a different question: "What are your desired outcomes? Six months from now, what specific changes do you want to see in your teams? What new actions will they be taking that they aren't today?"

After a moment, one of the leaders responded: "We need to increase contract conversions across the board. We are generating opportunities, but our closing rates are insufficient. In six months, I want to see a 25% conversion rate of analyses into contracts."

That revelation changed our approach entirely. Instead of developing a broad training program, we reframed it around their core business goal: improving conversion rates. We redesigned the program to focus on the initial analysis of opportunities and subsequently converting those analyses into contracts. The emphasis shifted from teaching general selling skills to honing the precise behaviors that would drive revenue growth.

This shift from "delivering training" to "driving transformation" significantly enhanced our credibility. We moved beyond simply fulfilling their initial request, and instead, we helped them achieve their key business objectives.

The lesson from that sales training meeting is universally applicable. When leaders move from delivery to transformation, they stop fulfilling requests and start owning outcomes.

At Hilti, product managers typically oversaw product training, adhering to a standard launch format and rarely requesting more than content distribution from the Learning and Development (L&D)

department. However, the launch of Nuron, a complete redesign of the battery platform powering Hilti's cordless tools, presented a unique situation.

Nuron represented more than just a product launch. It signified a fundamental shift in how customers would experience Hilti's entire battery-powered line. The new platform offered greater output, provided "smart" data for customers, and extended battery capability to tools previously requiring corded power, such as breakers. It also came at a higher price point and meant that existing tools would now be supported by legacy batteries.

This scenario transcended traditional product training; it presented a change management challenge that required buy-in from the sales team and trust from customers. Consequently, L&D was engaged early in the process.

Instead of overwhelming the sales force with a barrage of specifications and data, we implemented a scaffolded rollout: concise modules directly linked to real-world selling conversations. We cultivated excitement progressively, addressed potential resistance proactively, and equipped salespeople with compelling narratives to justify the higher cost.

The results were compelling:

- Market organizations requested accelerated rollouts due to the training's readiness, translatability, and tailored approach.
- Time to market and time to adoption were accelerated globally, providing Hilti with a significant competitive advantage.
- The sales goal was ambitious, with 70% of the sales force expected to sell the product line within two months. However, the structured rollout made it achievable.
- Most importantly, customers remained loyal despite the higher cost, as the sales force confidently demonstrated the new platform's value and the added data it unlocked.

BECOME A STRATEGIC PARTNER

This success was not the result of delivering training hours. It stemmed from a strategic partnership, aligning the rollout with Hilti's business goals. This role was earned through the pipeline conversion project and reinforced with the Evolution project. In the Evolution project, we fostered local and regional advocates while maintaining global alignment to ensure lasting change. Nuron built upon the credibility of these business-relevant solutions.

By demonstrating business acumen and strategic value, we transcended the role of training provider, helping the company transform. As a result, our executives consistently viewed L&D differently. We were no longer an afterthought but a trusted partner invited to participate in other strategic projects earlier in the process. Following these successes, stakeholders invited us in early, alongside sales and product leaders, to help shape strategy.

More broadly, product success is never solely about features; it is about adoption, trust, and change management. Companies that treat launches as mere training events often miss the larger opportunity. Conversely, companies that treat launches as transformations by connecting marketing, sales, operations, and learning build a competitive advantage.

Too many organizations still mistakenly equate shipping a feature with creating business value. Transformation occurs when every function aligns its work with measurable business outcomes and collaborates in tandem.

Business Acumen is the Foundation

Business acumen is the foundation that allows transformation to occur. I have long maintained that business acumen is the differentiator between functions that are consulted late and those that are invited early. Yet it is not sufficient for any functional head alone to possess this capability. Influence scales only when business understanding is embedded across the team.

When functional experts can speak credibly about margin pressure, customer churn, revenue growth, or operational risk, they shift how they are perceived. They move from being seen as implementers of requests to being partners in problem solving. Executives involve them earlier because trust is established through shared priorities and shared language. Business fluency signals that a leader understands what is at stake and can be relied upon to make trade-offs in service of enterprise goals.

CEOs do not fixate on activity counts. They worry about whether the organization has the leadership depth and execution capability required to deliver on strategy. Functional leaders earn credibility not by perfecting internal processes, but by demonstrating how their work strengthens the company's ability to grow, compete, and adapt. No executive wants to micromanage training catalogs, process maps, or operational details. They want confidence that capable leaders are in place and that initiatives are moving the business forward.

This distinction determines whether a function delivers outputs or drives transformation. While activities are necessary inputs, they are rarely sufficient evidence of value. When leaders focus exclusively on what was delivered rather than what changed, the connection to business outcomes remains unclear, and influence erodes.

In my experience, when CLOs remain anchored in training-centric thinking, three predictable consequences follow:

1. **Learning becomes commoditized.** When learning is framed primarily as courses and workshops, it is easily replaced by vendors, platforms, or automation. The function is reduced to delivery rather than differentiation.

2. **Workforce engagement declines.** Employees are motivated by progress, achievement, and problem solving. When training feels disconnected from real work and real outcomes, participation becomes compliance-driven rather than performance-driven. Executives notice the absence of behavioral change.

3. **Strategic influence diminishes.** Leaders who cannot articulate how learning contributes to growth, margin improvement, or risk mitigation are excluded from critical discussions. Over time, the function is narrowed to compliance or support activities rather than positioned as a driver of transformation.

My experience shows that each functional area has their own unfortunate consequences that occur when leaders become too activity focused. Business acumen changes this trajectory. It enables leaders to frame their work in terms that matter to the enterprise and to design initiatives that accelerate strategy rather than merely support it. In the next chapter, we will define business acumen more precisely and outline how to develop it across your team as a repeatable operating discipline, not an individual trait.

Reframing the Narrative

A transformative shift occurs when CLOs transition from training to driving organizational transformation. My work, including reframing sales training to focus on contract conversions and assisting Hilti in accelerating the global rollout of Nuron, demonstrates that learning can directly impact revenue, retention, and competitive advantage. While my Hilti examples are from L&D, this principle extends across all functions.

Consider the supply chain VP who transformed the perception of her department with a single presentation. Previously, her reports focused on on-time delivery percentages and logistics costs – accurate data points that failed to resonate with the board. However, in the face of a global supply chain disruption, she shifted her approach. Instead of presenting percentages, she highlighted how her team's proactive adjustments preserved a $50 million client relationship that was at risk. The same logistics data now conveyed a compelling narrative of protecting revenue and securing market share, elevating supply chain from a back-office function to a competitive differentiator.

This illustrates the power of reframing: by demonstrating how your work directly contributes to revenue generation, customer retention, or risk mitigation, you establish yourself as an essential asset rather than a mere overhead expense.

These implications are not merely theoretical; data substantiates their validity. Despite the increasing investment in learning and development, a significant gap persists between intent and impact. According to the 2024 LinkedIn Workplace Learning Report, only 8 percent of CEOs perceive a clear business impact from their L&D investments. This skepticism, held by over 90 percent of CEOs, is reflected in budgetary decisions and limited influence at the executive level. Gartner's 2025 HR Priorities Survey reinforces this concern, revealing that fewer than 20 percent of HR and L&D leaders are "highly confident" that their training strategies will achieve business goals. Leaders recognize that simply increasing training hours or introducing new courses is insufficient. However, when learning initiatives are directly linked to tangible outcomes, the results are substantial. RedThread Research has found that organizations that align their initiatives with business impact are 3.4 times more likely to report higher retention rates and 2.8 times more likely to financially outperform their competitors.

This challenge extends beyond learning. McKinsey research indicates that 70 percent of CEOs primarily evaluate marketing based on revenue growth and margin contribution, while only 35 percent of CMOs prioritize these metrics in their dashboards. This disconnect is further highlighted by Deloitte, which reports that up to 80 percent of CEOs lack trust in or are dissatisfied with the metrics presented by their CMOs unless those metrics directly correlate with pipeline, conversion, or revenue impact. While impressions, clicks, and lead counts may be relevant to marketing teams, they often fail to resonate with executive leadership.

Supply chain leaders face a similar credibility challenge. Recent industry insights reveal that 88 percent of executives still perceive the supply chain function as a cost center. However, data suggests otherwise: companies with high-performing supply chains are 79 percent more likely to achieve above-average revenue growth. The consequences of failing to demonstrate this connection are considerable, with the average supply chain disruption costing approximately $1.5 million per day, even for minor disruptions. CEOs view this not as an operational inconvenience but as a significant business threat. Supply chain leaders who can articulate their contributions in terms of revenue protection, resilience, and customer loyalty are more likely to transition from being viewed as operational managers to being recognized as strategic partners.

Across these functions, the overarching message remains consistent. Leaders who translate their impact into tangible business outcomes, such as organizational growth, cost savings, risk management, and shareholder value, become indispensable. Embracing a transformative approach enhances credibility, attracts resources, and secures a lasting position at the executive table.

PRACTICAL FRAMEWORK: TELL THE STORY

Transitioning from functional delivery to a transformative approach requires reframing how value is defined, designed, and communicated. The following framework can assist teams in making this shift systematically.

I. Start with Business Intent

Before initiating any project, first ask: What business transformation are we aiming to achieve? Translate requests (e.g., "We need sales training") into measurable outcomes (e.g., "We need to increase conversion by 25%"). Define the rationale in terms of growth, cost, risk mitigation, or innovation, reflecting the language used by your CEO or in shareholder reports.

2. Identify Key Levers of Influence

Map the aspects of your function that most significantly impact the desired outcome.

- L&D - Leader readiness, capability development, accelerated time-to-performance
- HR - Talent mobility, employee engagement, talent retention
- Operations - Cycle time reduction, quality enhancement, cost avoidance
- Marketing - Brand trust, conversion rate optimization, customer lifetime value maximization
- Sales - Customer retention, account development

This step clarifies the connection between your function and overall enterprise results.

3. Design for Behavioral Change, Not Just Activity

Translate the desired outcome into specific behavioral or process changes. Define how people will act differently upon successful implementation. Then, align your learning, systems, incentives, or communication interventions to enable those behaviors. This is where genuine transformation occurs.

4. Signal Value, Not Volume

Replace volume-based metrics with value-based metrics.

We will go deeper on the metrics and evidence leaders trust in Chapter 5. Tracking changes in work processes demonstrates the return on investment to leadership.

5. Communicate in Business-Centric Language

Frame updates and reports using the organization's strategic vocabulary. Instead of saying, "We delivered 20 modules," say, "We helped reduce time-to-adoption by 30%, accelerating revenue by X%." Emphasize outcomes, not outputs.

This framework anchors a transformation-oriented mindset in daily practice. Once leaders master this discipline, AI becomes a powerful tool to test assumptions, scan trends, and connect functional work to the enterprise narrative.

Each element of this framework strengthens credibility and influence:

- Define clarifies purpose and aligns with enterprise goals.
- Identify ensures focus on the most impactful levers.
- Design converts strategy into tangible behavioral change.
- Measure quantifies value, not volume.

"Tell the Story" reframes perception and fosters early inclusion in strategic conversations.

Over time, this becomes a self-reinforcing cycle of credibility: each iteration increases strategic influence. The more effectively you connect your function's work to tangible outcomes, the faster the organization recognizes you as a transformation partner, rather than a mere service provider.

~

LEVERAGING AI AS A STRATEGIC THOUGHT PARTNER

To maximize AI's value as a thought partner, provide it with the same information your executives are using.

Try this approach:

1. Compile 2–5 of the most recent CEO addresses (all-hands recordings, quarterly updates, or earnings calls).

2. Include relevant internal strategy documents (presentations, roadmaps, or memos).
3. Input these materials into AI, along with a concise description of the initiative you are working on. Then, pose the following question: "Based on these inputs, what are the top five business outcomes this company is prioritizing? And how could the [insert your function] contribute to each?"

For example, the responses for an L&D prompt might include:

- Accelerating customer acquisition → shorten new-hire onboarding.
- Expanding into new markets → Cultivate cross-cultural leadership capabilities.
- Improving margins → Enhance operational excellence training for frontline teams.
- Retaining top talent → Strengthen career development frameworks.
- Managing risk → Reinforce compliance and ethical decision-making protocols.

The point is not for AI to give you perfect answers. It is to push you and your team to align your narrative with the CEO's or the Executive Board's priorities and to anticipate risks the CEO may not have articulated yet.

When you start testing your initiatives against the words of your own executives, you stop being a functional leader and start becoming a transformation partner. This will also help you reframe your thinking, and it will enable your team members to supplement their existing knowledge of the organization's business goals. Remember, the narrative then becomes "In order to accelerate customer acquisition to reduce the effects of our competitors' market inroads, we will revise new hire onboarding to focus upon quickly preparing new hires to effectively focus upon the right clients with the right narratives." You are putting the business need at the center and making decisions to accomplish that goal every time you speak about that initiative.

~

TRANSLATION DIAGNOSTIC: IS YOUR NARRATIVE ENTERPRISE-READY?

Before you step into your next executive meeting, test how your story lands outside your function. Score each item 1 – 5, where:

1 = Rarely true | 3 = Sometimes true | 5 = Consistently true

Add your totals to see where you're strong and where you need to sharpen translation.

Section 1 – Activity vs. Consequence (The Activity Bias)

- I describe outcomes in terms of business value, not just completed tasks.
- Every major verb ("launched," "implemented," "trained") is followed by a business result ("which reduced X cost" or "accelerated Y revenue").
- My visuals and metrics emphasize the impact of work, not the volume of work.
- Stakeholders could restate my message without using my functional jargon.

Section 2 — Language Fit (The Language Mismatch)

- I use terms executives track on dashboards (revenue, margin, risk, productivity).
- Functional words like engagement, efficiency, or awareness are translated into enterprise equivalents (retention risk, cost savings, market reach).
- I explain assumptions behind metrics so non-experts grasp their significance.

- I preview what the business will gain or lose if action is or isn't taken.

Section 3 — Time Horizon (The Short-Term Framing)

- I connect today's results to future capability or cost avoidance.
- I report patterns and trajectories, not one-time achievements.
- I highlight how this work scales or compounds value over time.
- I identify what risks or dependencies remain, showing forward stewardship.

Total

3

CREATING BUSINESS ACUMEN

Upon assuming the role of Chief Innovation Officer at **ansr**source, I observed a critical area for improvement in our investment decision-making processes. Historically, projects were often funded based on the enthusiasm of their proponents. A leader's passion and persuasive abilities could propel an idea forward, even in the absence of a well-documented business case or a clearly defined target market. One example was the development of digital learning content to sell off the shelf. Significant investments were made. To make this content reflect the brand promise of subject matter quality, journalistic integrity, and educational expertise, the best subject matter experts were located, a team was built, and millions in set up costs, salaries, and marketing brought these incredible courses to life. However, this was not a company known for delivering off the shelf content. Their small client base hired them to develop custom content. Their sales and marketing staff did not know how to reach new clients for whom this would be of value. The content blurred the line between higher education and corporate at a time when those expectations were very different. Corporate wanted micro-learning content, but they wanted it highly applicable to their business needs and linked to behavior change or performance rather than knowledge. Higher education was focused on creating their own approaches, with some organizations trying to repurpose their instructor led content and others focused on building libraries, but not micro-content that they felt would devalue their expertise and the focus on longer certifications and degree programs. Unfortunately, this incredible content languished on the digital shelf, caught between vision, passion, and client interest.

While passion is a valuable asset, fostering creativity and commitment, it can be detrimental without a strategic filter. The company was allocating significant capital and resources to projects in industries characterized by low margins and declining demand.

To address these challenges, I implemented a structured approach. Each potential investment is now evaluated through a decision tree, requiring comprehensive answers to key questions:

- What is the proposed investment amount, and what is the expected duration?
- Who is the target market, and does it align with our core competencies?
- Are there external entities willing to provide funding or pay for the proposed solution?
- What is the probability of generating profitability in the short to medium term?

Only projects that successfully meet these criteria are considered for further investment.

This represented a fundamental shift in our approach. Rather than prioritizing ideas based on enthusiasm, we began allocating resources to initiatives with the greatest potential for creating business value. While other challenges remained, introducing this disciplined approach enhanced our business acumen as a critical filter.

Business acumen is not intended to stifle passion. Instead, it serves to channel it effectively. Leaders with a strong understanding of the business can align enthusiasm and creativity with strategy, profitability, and growth. This is the essential transition that functional leaders must make, shifting their focus from internal activities to enterprise-wide outcomes.

Business acumen is a term frequently used in leadership discussions but often lacks a clear definition. While financial fluency is a component, true business acumen encompasses a broader understanding.

At its core, business acumen is the ability to understand how an organization creates value and how decisions impact that value.

It involves viewing the business as a system:

- Understanding the organization's revenue generation model.
- Analyzing its competitive positioning and strategies for success in the market.
- Identifying customer needs and willingness to pay.
- Recognizing potential risks and opportunities.
- Applying these insights to one's functional area.

Leaders with business acumen possess knowledge beyond their specific function. They understand the interdependencies between departments and how the various functions of sales, operations, finance, technology, and customer experience collaborate to achieve organizational goals. They recognize the cascading effects of decisions made in one area on the entire business.

Support functions are particularly susceptible to credibility deficits when their narratives lack business acumen. Leaders in these roles risk appearing disconnected from executive priorities when they fixate on internal processes, activities, or metrics. For enabling leaders, the challenge lies not in directly generating revenue, but in cultivating trust that their function strategically influences revenue, risk mitigation, and growth initiatives.

- In HR, leaders may inadvertently prioritize metrics like "time to fill" or "employee engagement scores" without demonstrating their direct impact on strategy execution or talent readiness.
- In IT, teams often emphasize uptime percentages or system deployments, rather than articulating how technology facilitates faster innovation or enhances customer experience.
- In Procurement, the focus can remain narrowly on cost savings, potentially overlooking the broader implications for supply chain resilience or vendor-driven innovation.
- In L&D, training hours and satisfaction scores frequently take precedence over demonstrating how development initiatives accelerate overall business growth.

In each of these scenarios, the core issue remains consistent: leaders prioritize easily quantifiable metrics over the outcomes that executives value most. Without demonstrable business acumen, their contributions may appear disconnected from the organization's growth, profitability, and competitive advantage.

Consider framing your function's contributions through two distinct lenses: an internal and an external perspective.

- **The internal lens focuses on activities:** What was our requisition fulfillment rate? How many systems were deployed? How many employees completed training?
- **The external lens focuses on outcomes:** Did our hiring decisions facilitate successful expansion into a new market? Did our technology investments demonstrably reduce customer churn? Did our training program accelerate contract conversions?

Of course you should know your internal activities as these are signals that you are moving the needle. These are likely even the metrics in the targets that you give your teams. Activities can often be our leading indicators and can help us determine if we are likely to achieve the desired outcomes. The difference between being perceived as operationally competent and strategically indispensable often lies not in the work itself, but in how that work is framed.

HR Example

Internal lens

> We achieved a **92% requisition fulfillment rate**, filling **138 of 150 open roles** within an average of **41 days.**

External lens

> By prioritizing revenue-critical roles, we staffed **85% of customer-facing positions** in our new Midwest region within **30 days**, enabling the region to reach **$12.4M in pipeline generation** one quarter ahead of plan.

Why it matters

> Executives care less about *how many roles were filled* and more about *which roles were filled* and *what business outcome that enabled.* Speed only matters when tied to revenue or growth.

IT Example

Internal lens

> We successfully deployed **4 new systems**, completed **27 integrations**, and maintained **99.8% system uptime** across platforms.

External lens

The new CRM and billing integrations reduced customer onboarding time by **22%**, contributing to a **6.3% reduction in churn** among first-year customers and protecting approximately **$3.1M in annual recurring revenue.**

Why it matters

System uptime is table stakes. What earns credibility is showing how technology directly influenced customer behavior, retention, and revenue protection.

L&D Example

Internal lens

We delivered leadership training to **420 managers**, achieving a **94% completion rate** and an average satisfaction score of **4.6 out of 5.**

External lens

Managers who completed the program reduced regretted attrition by **9%** within six months, stabilizing teams in high-growth regions and avoiding an estimated **$2.7M in replacement and productivity costs.**

Why it matters

Completion rates and satisfaction scores describe effort. Executives want to know whether leadership capability translated into retention, stability, and financial impact.

Executives are primarily concerned with the external lens. They prioritize understanding how your function contributes to the company's growth, cost optimization, and competitive positioning. Lacking this perspective diminishes the impact and influence of your voice.

When support leaders demonstrate business acumen, they are viewed as strategic partners. Executives proactively include them in key discussions, confident that these leaders possess a comprehensive understanding of the business. This trust stems from their ability to transcend internal metrics and align their expertise with the organization's overarching objectives.

Therefore, business acumen serves as the foundational element for all other leadership competencies. Strategic talent development, cross-functional collaboration, and culture shaping are contingent upon a solid understanding of the core business, industry dynamics, and organizational context.

THE FUNCTIONAL TRAP: WHEN EXCELLENCE BECOMES IRRELEVANCE

Many functional leaders find themselves in what I refer to as the functional trap. This is a place where they are highly competent, deeply trusted within their domain, and yet increasingly peripheral to enterprise-level decision-making. The trap is insidious because it is built on behaviors that were once rewarded.

Early in a leader's career, excellence is often defined by three attributes:

- **Efficiency:** The ability to execute quickly, meet deadlines, and deliver work with minimal friction.
- **Responsiveness:** Being reliably available, solving problems as they arise, and saying "yes" more often than "no."
- **Technical mastery:** Deep expertise in one's function whether HR, L&D, IT, finance, or operations.

These qualities are valuable. In many organizations, they are precisely what lead to promotion into functional leadership roles. Teams run smoothly, stakeholders are satisfied, and issues are addressed before they escalate. The leader becomes known as dependable and capable and viewed as promotable.

The problem emerges when those same behaviors remain unchanged as expectations shift.

At the enterprise level, executives are not primarily seeking flawless execution within silos. They are seeking leaders who can make trade-offs, prioritize ruthlessly, and connect decisions to growth, margin, and long-term enterprise value. Efficiency without discernment becomes busyness. Responsiveness without strategic filtering becomes distraction. Technical mastery without contextual understanding of the affect on the larger business becomes narrow.

In my own career, I observed that leaders who continued to optimize for functional excellence were often unintentionally training executives to see them as service providers rather than strategic partners. They became exceptionally good at delivering what was asked and yet rarely influenced *what* should be asked in the first place. Over time, their proximity to decision-making decreased. This didn't happen because they lacked capability, but because their contributions were framed in operational terms rather than strategic ones.

This erosion of executive trust is rarely explicit. Leaders are not told they are being sidelined. Instead, they are simply included later, consulted after key decisions are made, or asked to "support" initiatives rather than shape them. Their work remains important, but their voice carries less weight.

One reason the functional trap persists is that it offers psychological safety. Function-specific metrics are familiar, defensible, and controllable. Completion rates, service levels, system uptime, engagement scores, or cost savings provide a sense of certainty and accomplishment. They also protect leaders from the discomfort of ambiguity and exposure. They positively impact your bonus or variable compensation and target achievement.

Enterprise-level conversations, by contrast, are inherently uncertain. They require leaders to speak about forecasts, probabilities, risks, and trade-offs often without perfect data. Initially this can feel nerve-wracking because it invites challenge, scrutiny, and disagreement. For many high-performing functional leaders, staying within the comfort of operational metrics feels safer than venturing into conversations about revenue impact, market dynamics, or capital allocation.

However, safety and relevance are not the same. Leaders who remain anchored in function-specific language may continue to perform well, but they gradually lose influence over the direction of the organization. Their work becomes transactional rather than transformational. Ironically, the very excellence that once propelled their careers can render them strategically invisible. Sometimes it happens because they step into enterprise conversations multiple times only to be shut down by higher leaders or their opinions discounted and they begin to question their own ability to see the bigger picture. Think of like when you were learning algebra or grammar; you didn't just give up because you got it wrong. Instead, you adapted, learning the rules and formulae that helped you complete the task well. Just because you might miss the mark initially does not mean that you should step back and stop trying.

Escaping the functional trap requires a conscious shift. Leaders must move beyond asking, *"Did we execute well?"* and begin asking, *"Did this materially change the trajectory of the business?"* That transition away from internal validation and toward enterprise impact is the foundation of true business acumen and a prerequisite for sustained executive credibility.

PRACTICAL FRAMEWORK: CULTIVATING BUSINESS ACUMEN

Business acumen requires deliberate cultivation, consistent reinforcement, and integration into team workflows. Without these measures, leaders and their teams may revert to familiar internal language and activity-based metrics, which often represent the immediate, day-to-day tasks and communication styles.

The following five practices can cultivate business acumen in daily work. While rooted in my experience leading an L&D team, these principles are equally applicable to any functional leader.

Require Financial Fluency

Team members learned to interpret basic financial reports and familiarize themselves with relevant terminology. Every leader should be able to understand a profit-and-loss statement. This does not require transforming HR managers into accountants or IT directors into CFOs, but rather ensuring they comprehend revenue, margin, and expenses. How does the business **make** money and where does it quietly **lose** money?

Over time, I found that financial fluency is less about mathematical precision and more about **pattern recognition.** Executives do not scan financials line by line. They look for signals that indicate enterprise health or risk. The numbers they consistently watch are not obscure: revenue growth and composition, gross and operating margins, cash flow, and exposure to risk. Together, these indicators tell a story about whether the organization is scaling sustainably or eroding value beneath the surface.

One of the most common missteps functional leaders make is equating financial success with short-term cost reduction. While expense control is important, aggressive cost cutting, particularly in areas like talent, technology, or capability development, can undermine future growth. A function may proudly report savings while unknowingly increasing attrition, slowing innovation, or weakening customer experience. In isolation, the numbers look positive. In context, they tell a very different story. I remember a story about a sales manager who extended the time between tire replacements on his fleet of cars. Short term reduction of expenses that increased the danger to the employees, the risk to the cars, and could have led to a much higher overall cost if it had not been rectified!

When my team reviewed the company's quarterly earnings alongside departmental figures, financial terms became demystified and integrated into our common lexicon. We asked simple but revealing questions: Where is margin expanding or contracting? Is revenue growth coming from new customers or deeper penetration of existing ones? Which costs are fixed, and which are investments intended to fuel future returns? What assumptions underpin our forecasts, and where are we most exposed if those assumptions prove incorrect?

Project budgets became more disciplined, not because we spent less, but because we spent with greater intentionality. Leaders learned to differentiate between expenses that merely sustained operations and investments that amplified enterprise value.

In one instance, this shift in financial fluency fundamentally altered a decision. We were evaluating whether to delay funding a capability-building initiative to meet a near-term cost target. Initially, the proposal appeared fiscally responsible. However, by examining margin trends and revenue concentration, it became clear that the initiative directly supported faster onboarding and reduced ramp-up time in a high-growth segment. Delaying it would have preserved short-term cash but constrained revenue capacity in the following quarters. With that context, the decision changed. The investment moved forward, not as a discretionary expense, but as a strategic lever tied to growth.

This is the essence of financial fluency; to understand how individual decisions ripple through the system. Leaders who develop this fluency are better equipped to engage in strategic dialogue, anticipate executive concerns, and make decisions that balance performance today with sustainability tomorrow.

In the context of amplified leadership, financial fluency enables sensemaking. It allows leaders to interpret signals, challenge assumptions, and connect their function's work to the broader health of the organization. Without it, leaders remain reactive, responding to financial outcomes after they occur. With it, they become proactive contributors to the enterprise's future trajectory.

Link Every Project To A Business Driver

Project approval became contingent on a clear connection to revenue growth, cost control, customer retention, or risk management. Each project lead was required to articulate this link concisely, in three minutes or less. Failure to do so necessitated project revision until the connection was evident. This process reinforced the importance of the discovery phase, where stakeholders defined their vision of success. We transitioned from fulfilling generic training requests or reactive "repair" orders to proactively driving targeted business changes.

This discipline fundamentally changed the quality of conversations across the organization. When leaders were required to anchor their requests to a business driver, vague problem statements surfaced quickly. Phrases such as "we need better communication" or "we should upskill managers" were no longer sufficient. Instead, stakeholders had to clarify what was actually at risk or opportunity in clear business terms such as missed revenue, delayed growth, rising costs, or exposure to regulatory or operational failure. In many cases, the original solution changed once the underlying business driver was made explicit. This shift reduced rework, prevented misaligned investments, and positioned my team as strategic partners who helped diagnose the business problem before prescribing an intervention. It also prepared these team members for future strategic positions because they understood how easy it was to fall back on functional terms and needs and how these had a knock-on effect on the broader business.

Simulate Executive Presentations

We practiced pitching initiatives as if presenting to the CEO or CFO, emphasizing the potential impact: "This will increase contract conversions, shortening the buying cycle and impacting revenue within the next quarter." "This will reduce onboarding time by 20%, accelerating revenue generation." "This will lower turnover costs, allowing sales managers to focus on revenue targets instead of candidate interviews." Initially, the exercises felt awkward. We struggled to substantiate our claims, stumbled over unfamiliar terminology, or lacked supporting metrics. However, with practice, confidence grew. When facing real executives, my team was **prepared to articulate the business value of our initiatives.**

More importantly, these simulations revealed gaps in our own thinking long before executives ever saw the work. We knew that a solution was not ready when it could not withstand executive-level questioning such as: *How will this affect margin? What assumptions are you making? What happens if adoption lags? What would you stop doing to fund this?* The exercise was less about presentation polish and more about strategic rigor. By rehearsing in this way, leaders learned to pressure-test their logic, anticipate trade-offs, and refine their narrative until it reflected enterprise reality rather than functional aspiration. Over time, this practice shifted mindsets: teams stopped preparing "updates" and started preparing decisions.

Recognize Business Outcomes

We shifted our focus from celebrating task completion or meeting deadlines to acknowledging tangible business results. Success was defined by a program's ability to shorten the sales cycle, improve retention, or strengthen the leadership pipeline – all key strategic goals for senior management. We maintained a laser focus on the intended outcomes of training programs and change initiatives, establishing leading and lagging indicators to gauge progress and identify key drivers of success. Instead of moving rapidly between projects, team members assumed ownership of rollouts and gathered post-implementation use cases.

This reframing also changed how leaders related to their work over time. When outcomes rather than activities became the point of reference, initiatives were no longer considered "done" at launch. Teams stayed engaged through adoption, behavior change, and business impact, developing a deeper sense of accountability for real-world results. This shift discouraged performative busyness and encouraged thoughtful follow-through. Leaders began asking different questions: *Who is actually using this? What decisions has it influenced? Where is it breaking down in practice?* By anchoring their work to outcomes rather than outputs, they learned to think beyond delivery and toward durability, which is an essential capability for leaders expected to influence enterprise performance.

Nurture Cross-Functional Peer Relationships

Each team member was expected to build deliberate, working relationships beyond our immediate function as a means of developing strategic literacy across the enterprise. Every function carries its own definition of risk and success. Finance worries about margin erosion, cash flow volatility, and forecasting accuracy. Sales is attuned to pipeline health, deal velocity, and competitive pressure. Operations focuses on throughput, quality, and scalability. Sustainability, compliance, and legal functions monitor reputational exposure and regulatory consequences. Leaders who fail to understand these distinct risk languages often struggle to gain influence because they are framed in terms that do not resonate with executive concerns.

As team members learned to interpret and speak these different functional languages, their influence increased markedly. Conversations became more precise. Proposals were positioned in terms of trade-offs and downstream impact across the system. The cross-functional relationships provide fluency that builds credibility quickly; leaders were demonstrating an enterprise mindset. Over time, these relationships shifted how team members were perceived.

A useful reflection question emerged from this practice: *If you were removed from the organization tomorrow, which functions would feel the impact and why?* Leaders who could answer this clearly tended to have broader influence, stronger judgment, and greater readiness for expanded roles. By contrast, leaders whose impact was confined to their own function often struggled to articulate their value in enterprise terms. Developing cross-functional strategic literacy transforms relationship-building from a "soft" skill into a critical leadership capability that directly affects a leader's ability to shape decisions, manage risk, and contribute meaningfully to long-term organizational performance.

These practices, while straightforward, collectively cultivate a culture where business acumen serves as the lens through which executives make decisions and measure success.

THE CONSEQUENCES OF LACKING BUSINESS ACUMEN

Consider José Alvarez at Tucanes, the 2,500-person company introduced in Chapter 1. During his board presentation, José narrowly focused on training hours delivered and requested an additional headcount to manage leadership training. His request was denied, ultimately leading to his demotion. However, the true loss extended beyond José's personal credibility; it represented a missed opportunity for the company. Had José approached the conversation with business acumen, he could have completely reframed his request, positioning L&D as a catalyst for growth and transformation. For example:

- **AI enablement:** Recognizing the C-Suite's concerns about competitors reducing costs through AI, José could have proposed a dedicated team to facilitate the ethical and productive adoption of AI tools across the workforce, accelerating performance to maintain competitiveness and achieve ambitious targets.

- **Operational efficiency:** Given the increasing emphasis on cost containment, José could have proposed outsourcing low-value tasks, such as training administration and LMS support, to free up budget for higher-impact initiatives like targeted sales training designed to increase share of wallet with existing clients.

- **Strategic partnership:** José could have positioned himself as a strategic partner capable of addressing challenges arising from industry changes. He could have offered to create a dedicated team within L&D to collaborate directly with executives on issues such as attrition, customer retention, and growth challenges.

Instead of appearing short-sighted and operational, José could have presented himself as forward-thinking and strategic. This approach could have secured him both budget and greater trust from the CEO, paving the way for increased influence. By setting modest goals, he reinforced the perception of L&D as a service function. Conversely, by aiming high, he could have reshaped the entire conversation. This scenario frequently occurs in support roles, but it can affect any function when individuals become internally focused and complacent.

This assertion is supported by global research, which confirms that business acumen is a defining characteristic of strategic leadership across industries and functions. A 2025 study by Culture Plus revealed that professionals with strong business acumen are five times more likely to be promoted into leadership roles. This trend indicates that advancement stems not only from technical expertise but also from a comprehensive understanding of business operations and the ability to align individual contributions with strategic outcomes. Leaders who demonstrate strategic skills experience accelerated career progression because executives perceive them as capable of driving enterprise-wide results rather than merely achieving departmental objectives.

Chief People Officers concur. In a September 2025 World Economic Forum outlook, 100% of CPOs ranked business acumen among their top three success factors, with nearly 90% identifying it as their single highest priority. As one CPO stated, "The separation between people and business is no longer viable." This principle applies across all functions. Whether leading sales, product development, finance, or operations, the capacity to connect one's work to growth, margin, and long-term enterprise value is now a prerequisite for establishing credibility.

However, many professionals still lack proficiency in this area. The 2025 Pulse of the Profession report by PMI disclosed that only 18% of project professionals demonstrate strong business acumen, despite senior leaders overwhelmingly recognizing its necessity for driving organizational value. This suggests a sobering implication: many initiatives fail to achieve their intended impact due to leaders' inability to frame or manage them within a business context.

The message is clear: leaders who cultivate business acumen gain trust earlier, advance more rapidly, and are included in strategic decision-making processes. Conversely, those who do not demonstrate business acumen remain limited to tactical execution, while others shape the future of the business. At Hilti, Andrea Moretto, as Head of L&D for a southern European region, exemplified this by evolving into an astute strategic business partner. His knowledge of the business and his ability to link learning topics to business objectives earned him a trusted position at the table. He made deliberate, business-oriented decisions regarding resource allocation, consistently reinforcing the importance of focusing on business targets and outcomes, thereby leading by example for both his team members and peers.

$\sim$

LEVERAGING AI AS A STRATEGIC THOUGHT PARTNER

AI can serve as a powerful thought partner in the development of business acumen by surfacing the kinds of questions seasoned executives instinctively ask. Used well, AI increases executive readiness by helping leaders pressure-test their thinking before it is exposed in high-stakes forums such as board meetings, budget reviews, or investment discussions. Used poorly, it does the opposite. AI does not replace business acumen; it exposes its absence. Leaders who rely on

AI to compensate for weak strategic framing or unclear business logic will find those gaps amplified, not concealed.

The most effective use of AI is therefore diagnostic rather than performative. Leaders can use it to challenge assumptions, identify blind spots, and stress-test the coherence of their narrative in advance. For example:

1. **Input Core Information.** Provide a concise description of a current team initiative, such as a training program, system upgrade, new benefits process, sales campaign, or vendor change.
2. **Add Recent Priorities.** Input the description into AI along with your company's most recent strategic priorities.
3. **Select a Prompt Role.** Prompt AI with the following: "Assume the role of the CFO preparing for a board meeting. What three questions would you ask me about this initiative to assess its alignment with our priorities and its potential to improve profitability, growth, or risk management?"
4. **Ask for Foresight.** Follow up with: "What external market or competitive risks might affect these priorities that I should anticipate in my plan?"

The value of this exercise lies not in the answers AI provides, but in the discipline it enforces. The questions often reveal gaps in logic, untested assumptions, or missing financial and market context. Leaders can then refine their thinking, strengthen their rationale, and align their proposals more closely with enterprise realities. Practiced consistently, this approach trains leaders and teams to think the way executives do by anticipating scrutiny, weighing trade-offs, and framing decisions in terms of growth, margin, and risk.

A strong foundation in business acumen remains essential. AI cannot substitute for judgment, context, or experience. However, when used intentionally, it becomes a mirror—reflecting whether a leader is prepared to engage at the executive level or still operating from a functional perspective. Leaders who use AI in this way arrive at critical conversations not in a reactive posture, but as informed, credible peers capable of discussing enterprise value, competitive advantage, and long-term sustainability with confidence.

A strong foundation in business acumen is essential. Without it, you will constantly be in a reactive position. With it, you can engage with other executives as an equal and discuss growth, profitability, and competitive advantage.

4

THE LEADERSHIP CAPACITY RISK

Developing adaptable, cross-functional leaders expands your sphere of influence in ways no single role or title ever can. When leaders invest in others, they create continuity with people who can carry forward decisions, values, and strategic intent when the leader is not present. This is not an abdication of authority, nor a dilution of control. It is the mechanism through which organizations sustain performance over time. Leadership capacity determines whether strategy survives leadership transitions, market shifts, and inevitable disruption.

Yet leadership capacity is one of the most underestimated risks in modern organizations. Executives routinely invest millions in strategy, technology, and transformation initiatives while assuming that leadership capability will somehow keep pace. When it does not, the gap becomes visible only after momentum stalls. Missed handoffs, inconsistent execution, and fragile succession pipelines are rarely labeled as leadership failures. They are more often misdiagnosed as execution issues, resistance to change, or cultural problems. In reality, they are symptoms of insufficient leadership depth.

Leaders who possess strong business acumen but fail to develop adaptable, empathetic successors ultimately constrain both their organization's future and their own. I witnessed this contrast clearly when I joined Hilti, where two leaders who were both intelligent and well-positioned took markedly different approaches to developing people, with equally different outcomes.

One leader, whom I will call Karen, led the global training development function. I was her first U.S.-based hire, brought in as a project manager for sales training. From the outset, she discouraged interaction with the North America training team and physically located me in a separate building. Her rationale was that local perspectives might "color my opinions" and compromise the global view. Engagement across departments was not encouraged; control and consistency were prioritized over connection and context.

As a result, I learned independently. I explored the intranet, sought out documents, and asked deliberate questions of people I encountered through projects. My prior experience had taught me that I could not add value without understanding the business—its customers, its competitive environment, and its strategic priorities. Time, I believed, was too valuable to spend operating in isolation. Karen's leadership approach was unmistakable: information was managed tightly, relationships were constrained, and development occurred, if at all, through formal structures rather than human connection. She led with authority, but not with trust. Ultimately, she left the organization under unfavorable circumstances, remembered less for what she built than for what failed to endure. No meaningful talent pipeline followed her tenure.

In contrast, Leo McKnight led training for Hilti North America. I first met him by chance on a flight to Europe. His background was firmly rooted in the business of sales, sales management, and sales leadership. His credibility came not from positional authority, but from lived experience in the core of the enterprise. What stood out most, however, was his humility. Leo introduced himself with ease, creating immediate psychological safety. He never assumed he had all the answers, nor did he position himself as the sole expert.

Leo was a natural developer of people. He recognized contributions openly, asked thoughtful questions, and entrusted others with responsibility earlier than most leaders would. Under his leadership, training was no longer perceived as a service function operating on the margins of the business. It became a strategic enabler, directly influencing performance, readiness, and growth. Through Leo, I came to see that my move into L&D was not a step away from the business, but a way to influence outcomes at greater scale. The trajectory of his team reflected that belief. Trainers who worked under his guidance progressed into roles such as sales directors, general managers, heads of strategic marketing, and global leaders. What they carried forward was not just technical expertise, but a shared mindset of business acumen paired with a deep commitment to developing others.

Two leaders. Two distinct approaches. One centralized knowledge and constrained growth. The other invested in people and built capacity far beyond himself. Only one left a legacy that extended beyond their tenure.

This contrast reveals a fundamental truth: organizations do not scale through strategy alone. They scale through people who can interpret strategy, adapt it to context, and execute it consistently. Individual performance, no matter how strong, is insufficient. What matters is organizational depth in the number of leaders who can think, decide, and act with enterprise perspective under pressure.

FROM INDIVIDUAL PERFORMANCE TO ORGANIZATIONAL DEPTH

Many leaders are promoted because they are exceptional individual contributors. They solve problems quickly, make sound decisions, and deliver results. Early success reinforces the belief that personal excellence is the primary path to influence. Over time, however, this model reaches its limit. As scope expands, the leader's effectiveness becomes constrained not by their own capability, but by the capability of those around them.

Organizational depth is the antidote to this constraint. It is built when leaders shift their focus from *what they can personally deliver* to *what the organization can sustain without them*. This requires a deliberate transition from optimizing individual performance to designing systems, roles, and experiences that develop others. Leaders who fail to make this shift often become bottlenecks, inadvertently centralizing decisions and slowing execution. Those who succeed create leverage. Their impact compounds because leadership capacity exists at multiple levels, across functions, and in moments of uncertainty.

The remainder of this chapter examines how leaders can intentionally build that depth regardless of function or title and why doing so is no longer optional in environments defined by speed, complexity, and constant change.

Developing Talent Is Not Optional

Leo's impact was not accidental, nor was Karen's outcome unusual. Both were products of how they understood leadership responsibility. One viewed development as central to enterprise health; the other treated it as peripheral. This distinction matters because leadership capacity does not emerge organically. It is built or neglected by the choices leaders make every day. Leo's legacy at Hilti underscores a fundamental truth that extends far beyond the realm of L&D: developing people is not the sole responsibility of HR or training leaders. It is the obligation of every leader, in every function. No strategy can succeed without capable individuals to execute it. A meticulously crafted financial plan, ambitious sales targets, or an innovative product roadmap are all rendered ineffective without leaders who can drive implementation.

In the past, organizations could afford a more gradual approach to leadership development, allowing promising employees to progress over time through experience and on-the-job learning. However, this is no longer a viable option. The rapid pace of change, driven by technological disruption, demographic shifts, and increasingly agile competitors, demands leaders who can quickly adapt, operate across functions, and apply their diverse experiences to novel situations.

The true value of talent development lies in its ability to cultivate cross-functional leaders who possess a holistic understanding of the business system. While technical experts can be developed through targeted education, training, and experience, organizations often lack effective and scalable strategies to develop:

- A finance manager who can rotate into operations and understand customer impact.
- An IT leader who can lead a transformation project in sales or marketing.
- A procurement expert who can step into sustainability or supply chain resilience.
- A training professional who can speak confidently about attrition, productivity, and growth.

Leaders who prioritize talent development in this manner view filling current roles as a strategic step toward building the C-suite of the future. Some leaders may hesitate, questioning the wisdom of preparing their own replacements or making the path easier for others.

The answer is clear: neglecting talent development harms the company and diminishes your influence. Leaders who hoard knowledge limit organizational growth, while those who develop others extend their impact far beyond themselves. Their ideas, values, and vision extend far beyond their individual reach. Investing in talent development yields three significant benefits:

1. **Amplified Influence:** Developing leaders who share your mindset ensures your ideas and values are represented even when you are not present.
2. **Enhanced Team Performance:** Developed and challenged individuals take ownership, proactively solve problems, and free you to focus on strategic priorities.

3. **Enduring Legacy:** Cultivating adaptable, cross-functional leaders ensures you are remembered for both your accomplishments and the people you developed, creating a legacy that transcends individual projects or budgets.

Developing people is not relinquishing power; it is expanding your reach. Talent development is ultimately a capacity decision: how much leadership depth the organization can draw on when pressure increases.

What Leaders Who Build Capacity Do Differently

Leaders who successfully build leadership capacity do not rely on formal programs or episodic development efforts. They operate differently on a daily basis. Their choices with what work they assign, how they evaluate success, and where they spend their attention systematically increase the organization's ability to think and act without them. Capacity is not created through intention alone; it is shaped through repeated leadership behaviors that signal what the organization values and rewards.

First, these leaders spot potential under pressure, not in comfort. They pay close attention to how individuals respond when the path is unclear, stakes are high, or support is limited. Rather than reserving challenging work for proven performers, they deliberately assign stretch responsibilities to those who demonstrate curiosity, adaptability, and ownership. These assignments are not reckless; they are paired with context, access, and feedback. The goal is not to test loyalty or endurance, but to observe judgment, learning agility, and resilience, all of which are qualities that predict enterprise readiness far more accurately than tenure or technical expertise.

Second, leaders who build capacity design cross-functional exposure rather than protecting silos. They recognize that leadership capability expands fastest when individuals are required to navigate unfamiliar perspectives, constraints, and priorities. Rotations, project-based assignments, and cross-functional problem-solving are used intentionally, not as career rewards but as development accelerators. Through this exposure, emerging leaders learn to translate ideas across functions, anticipate downstream impacts, and make trade-offs that reflect enterprise reality rather than functional preference. Over time, this reduces coordination friction and improves decision quality across the organization.

Third, these leaders balance accountability and empathy as a performance system, not a personality trait. They set clear expectations and hold people responsible for outcomes, while also recognizing that growth occurs under conditions of trust. Accountability without empathy breeds fear and risk aversion; empathy without accountability erodes standards and momentum. Leaders who build capacity navigate this tension deliberately. They provide candid feedback, address performance gaps early, and support individuals through challenge without lowering the bar. This balance encourages ownership and psychological safety simultaneously; both essential for developing leaders who can operate independently under pressure.

Fourth, leaders who build capacity scale through advocates rather than control. They resist the instinct to centralize expertise or decision-making, even when they are capable of doing so more quickly themselves. Instead, they identify and empower local champions who can contextualize strategy, reinforce standards, and influence peers. These advocates extend leadership reach, accelerate adoption, and create redundancy that strengthens resilience. Control may feel efficient in the short term, but it constrains capacity. Advocacy builds alignment without dependency.

Finally, leaders who build capacity normalize succession as continuity planning, not personal replacement. They speak openly about future roles, readiness, and development paths. They invest in preparing others for responsibilities they may never formally hold, understanding that readiness often precedes opportunity. Rather than fearing obsolescence, these leaders recognize that organizations penalize single points of failure. By developing successors, they reduce risk, increase flexibility, and reinforce a culture where growth is expected and supported. Their influence endures because it is embedded in people, not positions.

Taken together, these behaviors shift leadership from individual excellence to organizational strength. Leaders who build capacity are not merely effective in their own roles; they increase the organization's ability to adapt, decide, and perform over time. Their impact is measured not only by what they personally deliver, but by how well the organization functions in their absence. That distinction is what separates capable leaders from truly scalable ones.

The Compounding Effect of Leadership Capacity

Leadership capacity compounds in ways that are often underestimated because its impact unfolds over time rather than in immediate results. Unlike discrete initiatives that deliver short-term gains, leadership capacity strengthens the organization's ability to respond, adapt, and execute repeatedly. Each capable leader developed increases the organization's decision-making bandwidth, reduces dependency on a small number of individuals, and improves consistency under pressure. Over time, these effects accumulate, creating an organization that can absorb change without losing momentum.

Organizations with strong leadership capacity move faster with fewer errors. Decisions are made closer to the work by leaders who understand both local context and enterprise priorities. Escalations decrease not because problems disappear, but because leaders are equipped to resolve them. This reduces friction, shortens response times, and allows senior leaders to focus on strategic direction rather than operational intervention. Speed becomes a structural advantage rather than a function of heroics.

Leadership capacity also lowers risk in less visible but equally important ways. When leadership depth exists across functions and levels, the organization becomes less vulnerable to turnover, market shifts, or sudden changes in direction. Transitions occur with greater stability. Institutional knowledge is retained and applied rather than lost. Culture remains coherent because values are carried by people, not policies. The organization does not rely on a small group of individuals to hold everything together.

Financial performance reflects this compounding effect. Companies with strong internal leadership pipelines consistently outperform those that depend heavily on external hiring for critical roles. Internally developed leaders reach effectiveness faster, require less acclimation, and make better decisions earlier because they understand the business system. The cost savings associated with reduced recruitment, onboarding, and early attrition are meaningful, but the larger benefit lies in sustained execution and continuity during periods of growth or disruption.

Over time, leadership capacity becomes self-reinforcing. Capable leaders attract and retain high-potential talent because growth is visible and credible. Expectations remain high because accountability is distributed rather than concentrated. Learning accelerates because experience is shared and reflected upon across the organization. Each generation of leaders raises the baseline for the next, creating momentum that is difficult for competitors to replicate.

The absence of leadership capacity produces the opposite effect. Organizations become brittle. Decision quality declines under pressure. High performers burn out or leave due to overload. External hires struggle to integrate and often depart before contributing meaningfully. The organization expends increasing energy maintaining stability rather than advancing strategy. These outcomes rarely appear in isolation. They compound as well, but in the wrong direction.

Leadership capacity is therefore not a soft outcome or a long-term aspiration. It is a structural asset that shapes performance over time. Leaders who invest in developing others do not simply improve their teams. They change the trajectory of the organization. The return on that investment grows with each decision made well, each transition navigated smoothly, and each leader prepared to step forward when needed.

OPERATING PRACTICES THAT BUILD LEADERS

Here are five strategies for leaders to integrate talent development into their daily work. I am reminded of Andreas Markgraf, an early learning business partner I worked with. Drawing from his experience in Sales and Corporate Excellence, he quickly recognized learning as a performance driver. Witnessing his ability to connect development strategy to sales outcomes solidified his conviction. Andreas went on to lead major business units and became a vocal advocate for people development within Hilti. His continued growth and support of others, even after our paths diverged, exemplifies why developing talent is a core leadership responsibility. Andreas did not simply progress individually; his trajectory reflected how early investment in cross-functional capability creates leaders who can scale impact across the enterprise.

The behaviors described earlier become durable only when they are translated into consistent operating practices. Leadership capacity is not built through isolated moments of coaching, but through how leaders assign work, distribute responsibility, and make development visible in everyday decisions. The following practices reflect how leaders embed development into the normal flow of work, rather than treating it as a separate activity.

- **Identify Potential Early:** Talent often emerges during challenging assignments. Provide opportunities that slightly exceed individuals' current skill sets and observe who embraces the challenge, adapts, and learns. When assigning a stretch goal, observe their response: Do they take ownership and adapt readily? During my time at Hilti, I observed future leaders distinguish themselves by thriving under pressure in such situations.
- **Facilitate Cross-Functional Exposure:** Avoid confining talent within a single department. Rotate individuals through projects in different areas and pair them with mentors from outside their function. Cross-functional

experience fosters resilience and broadens perspectives. Leaders develop more rapidly when they gain insights beyond their immediate domain. This exposure enhances adaptability and often reveals hidden strengths, while also strengthening their professional network.

- **Model Accountability and Empathy:** Effective leaders balance high expectations with genuine care. They challenge individuals to achieve results while providing support as individuals. Accountability without empathy breeds fear, while empathy without accountability fosters complacency. Striking the right balance promotes growth. At Hilti, successful leaders expected results while acknowledging the impact of personal lives on work performance. This balance fostered both growth and loyalty.

- **Cultivate Advocates and Local Champions.** When implementing a new sales approach at Hilti, we transitioned to a globally modified sales force dynamic. Instead of relying solely on headquarters-led training, we identified and empowered advocates in local and regional markets. These champions delivered contextualized training, supported by global resources. This approach fostered smoother adoption, faster alignment, and a more robust leadership pipeline.

- **Normalize Succession Planning.** Treat succession planning as a transparent process, openly discussing the development of future leaders. Preparing successors is not a sign of weakness, but rather the hallmark of a strong leader who prioritizes continuity and growth. Ascertain individual aspirations, provide developmental opportunities, and offer constructive feedback. Avoid hindering growth by blocking advancement or setting individuals up for failure. Recognize that individuals may not be fully prepared for every role initially; allow room for professional development.

THE STAKES OF LEADERSHIP CAPACITY

Without deliberate talent development:

- Organizations stall when key leaders depart.
- Core values and culture erode as external hires struggle to integrate.
- Companies incur prohibitive costs to acquire external talent.

With deliberate talent development:

- Organizations adapt quickly to market change.
- Leaders perpetuate the vision and values of today into tomorrow.
- Companies build resilience, agility, and long-term success.

The data underscores a critical imperative for leaders: companies that prioritize strategic talent development consistently outperform those that treat it as an afterthought. According to the 2024 DDI Global Leadership Forecast, organizations with robust leadership development systems were 2.4 times more likely to meet their performance targets and 3.5 times more likely to financially outperform their peers. Robust talent pipelines are essential and directly correlate with market success.

For instance, McKinsey's 2025 analysis revealed that companies filling 60 percent or more of senior roles internally, rather than relying on external hires, were twice as likely to deliver above-average shareholder returns. Developing internal leaders strengthens culture and continuity and enhances financial performance.

Korn Ferry's 2025 research confirmed that internal promotions saved organizations 20 to 30 percent compared to external hires, with internally promoted individuals demonstrating a significantly higher likelihood of long-term retention. High-potential employees not only perform effectively but also exhibit loyalty when they perceive genuine investment in their growth.

I have observed these principles in practice. A consumer goods company I advised invested significantly in succession planning following a wave of retirements that threatened leadership gaps. Resisting the temptation to immediately recruit externally, they placed mid-level managers in accelerated leadership rotations. Within 18 months, they had cultivated a strong leadership pipeline. As competitors struggled with costly external hires, this company surged ahead with leaders who possessed deep knowledge of the business and its culture.

In contrast, a technology services firm neglected talent development for years, operating under the assumption that leadership could be acquired externally as needed. However, when several senior executives departed simultaneously, the ensuing scramble to recruit replacements incurred substantial costs in recruiter fees and onboarding expenses. More critically, the majority of external hires struggled to integrate into the company culture, resulting in high turnover at the leadership level. This instability cascaded down to frontline staff, negatively impacting employee morale and productivity. The firm not only experienced financial losses but also suffered reputational damage, eroding trust with investors and clients.

These findings underscore a critical point: talent development is not merely a peripheral activity but a measurable driver of business advantage. Organizations that prioritize leadership development achieve their strategic objectives more efficiently, retain top talent, and cultivate resilience for long-term success.

For C-suite executives, this perspective should inform how they allocate their time and resources. Investing time in coaching emerging leaders, developing robust succession plans, and providing managers with opportunities for growth should not be viewed as discretionary. Neglecting these responsibilities carries significant consequences, undermining future performance and necessitating the acquisition of talent at a premium.

~

LEVERAGING AI AS A STRATEGIC THOUGHT PARTNER

AI can strengthen leadership capacity only when it is used to sharpen judgment rather than replace it. In the context of talent development, its greatest value lies in helping leaders think more rigorously about readiness, risk, and future capability requirements. AI does not develop leaders. It reveals where leadership capacity is strong, where it is fragile, and where assumptions have gone untested.

Used intentionally, AI allows leaders to step out of day-to-day operational bias and view their talent pipeline through an enterprise lens. It can surface patterns that are difficult to see from within a single function and prompt questions that senior executives and boards are increasingly asking about leadership depth and succession. This makes AI particularly effective as a preparation tool, enabling leaders to pressure-test their thinking before leadership gaps are exposed through disruption, attrition, or failed execution.

One practical approach is to use AI as a diagnostic partner:

1. **Gather Documents.** Compile your organization's most recent strategic documents, investor communications, or CEO addresses.
2. **List Priorities.** Identify the key priorities for the next two to three years, particularly those that require cross-functional execution or significant change.
3. **Check For Needed Capabilities.** Use the following AI prompt: "Given these priorities, what leadership capabilities will be most critical to achieve these priorities?"
4. **Ask AI to Coach You.** Follow up with: " What five questions should I ask my high-potential leaders to assess their readiness to manage ambiguity and make enterprise-level decisions?"

The power of this exercise is not in the answers AI provides, but in the reflection it provokes. The questions often expose misalignment between strategy and development focus, reveal over-reliance on a small number of leaders, or highlight gaps that could become constraints under pressure. These insights give leaders the opportunity to act early, adjusting assignments and development priorities before risk materializes.

AI must be used with discipline. It cannot substitute for contextual understanding or human judgment. When leaders treat AI outputs as definitive, they weaken rather than strengthen leadership capacity. When they treat AI as a catalyst for deeper dialogue and shared sense-making, it becomes a valuable tool for building a more resilient organization.

Leaders who use AI in this way signal a clear expectation. Leadership readiness is not assumed. It is examined, challenged, and developed intentionally. In doing so, they reinforce a culture where growth is aligned with strategy, succession is proactive rather than reactive, and leadership capacity continues to deepen over time.

5

COMMUNICATING RESULTS EFFECTIVELY

Effective leaders possess a strong understanding of the business, cultivate their teams, and, crucially, communicate results effectively. These leaders are the driving force behind transformative change.

The boardroom was silent save for the hum of the projector. Denise, a learning and development leader in a manufacturing company, stood ready to present. After weeks of intensive preparation, she had compiled her team's accomplishments, analyzed data from the learning management system (LMS), and validated key metrics. She was eager to showcase the team's increased productivity, which aligned with the company's overall growth. While her efforts were commendable and her data thorough, executive audiences prioritize impact over effort. The intention here is not to criticize Denise, but to highlight how subtle narrative choices can determine whether a function is adequately funded or marginalized.

The first slide displayed the following:

- "18,000 training hours delivered."
- "92% completion rate."
- "4.6 out of 5 satisfaction score."

She turned to the executives. "We've had a very productive year, and enrollments in all of our classes are up." In that moment, the executive brain does what it always does: it scans for growth, cost, risk, and value. Hours, completions, and satisfaction don't map cleanly to any of those. Without translation, leaders unconsciously apply the *"so what?"* filter and move on.

The CFO leaned back in his chair, arms crossed. "That's fine, Denise. But tell me, how did all of these training hours improve revenue? Did you train people on how to reduce our operating costs? Did these trainings make us more competitive?" This is the point where the story must pivot from *what we did* to *what it changed*. A single sentence can turn the room: "In plants where we delivered the program, time-to-competency dropped 21%, which pulled $3.1M in productivity forward and reduced scrap by 9%."

Denise froze. She hadn't expected these questions. She fumbled to the next slide, but the numbers there were more of the same. Lack of access to commercial or operational dashboards is real. It's also solvable, and it's your job to solve it. When you don't have the data, narrate the path to it: "We don't yet have customer or cost files tied to this cohort. Here's the pull we've requested from Sales and Finance, and here's the proxy we used this quarter (cycle time and rework). We'll report the full business impact next review."

She began to explain that they did not have access to the customer dashboard or the revenue files. She expressed how responsive they were when asked by various departments to deliver training programs and how they had done more training this year with the same headcount and budget, but that she wanted to expand their offering so they could support more departments. She even mentioned that HR had stated that a common question during interviews was how much personal development the company offered.

The room stayed silent. Finally, the CEO broke in, thanking her for the update and moving the agenda forward. Silence is a signal. The CFO heard unfunded activity and margin risk. The COO heard no change to throughput or quality. The CEO heard a claim on attention with no return on strategy. No one questioned Denise's effort; they questioned its consequence.

As Denise packed up her laptop, she felt the slow burn of embarrassment and frustration. She had worked so hard and still missed what mattered. That realization, though painful, is what eventually turns practitioners into leaders. That presentation didn't just end her request for additional budget; it cemented a perception: learning was a cost, not an investment. Denise's function delivered activity, not results.

A few months later, we encounter another company with a different L&D leader, Marcus. In contrast to Denise's story, which ended in silence, Marcus's begins with intention. He redesigned not only his slides but also the narrative he presented. Each data point in his presentation was structured to preemptively address executive questions. Marcus stood confidently before his executive team, his slides reflecting a deliberate focus on specific executive triggers. He considered the COO's interest in time and productivity, the CFO's focus on cost and cash impact, and the CEO's concern for growth and margin expansion.

Marcus wasn't merely reporting results; he was anticipating decision-makers' perspectives.

- New-hire onboarding reduced from 12 weeks to 8, with the time to payback shortened by an additional 2 weeks, resulting in 6 weeks of productivity per new hire.
- Average time to productivity shortened by 33%, equating to $4.5M in accelerated revenue capture.
- Retention among first-year employees improved by 18%, saving $2.2M in turnover costs.

His narrative followed a clear progression from baseline to improvement to business impact. He began with time, transitioned to money, and concluded with people. The sequence was deliberate, mirroring the order of executive priorities: speed, profit, stability. Marcus didn't focus on hours of training; he emphasized time, money, growth, and retention. These were the same outcomes the executives were measuring in other areas of the business.

When the data aligns with the narrative executives want to hear, convincing them becomes unnecessary. The atmosphere in the room shifted from evaluation to collaboration, and the reaction was immediate. The COO leaned forward and asked, "If you can achieve this in onboarding, can the same approach be applied to sales training?"

The CFO nodded, adding, "If you can achieve cost reductions like that, we should double your budget." The CFO's statement wasn't an act of generosity; it was a rational assessment. When a function demonstrates clear financial leverage, the budget transforms into investment capital rather than overhead.

Marcus left the meeting with approval and momentum, relieved that his work resonated with the company's strategic objectives. Confidence arises when effort aligns with understanding. The power of Marcus's narrative wasn't in larger numbers but in a different tone. His voice conveyed ownership of business outcomes, not just stewardship of a department. This tonal shift signaled partnership, positioning his team as a strategic partner delivering measurable results.

The difference between Denise and Marcus wasn't talent or effort but language. One spoke in learning metrics, while the other spoke in results that executives valued. As Kevin Yates, the L&D Detective, often reminds us, activity metrics are not evidence of impact but rather clues. Unless leaders connect these clues to tangible business outcomes, executives will perceive training as noise rather than a signal.

Marcus's approach follows a repeatable narrative pattern:

1. Start with the problem or cost of inaction.
2. Show the shift or intervention.
3. Quantify the business consequence.
4. Conclude with an invitation to explore further applications: "How might we apply these insights in other areas?"

This approach transforms data into a valuable dialogue, fostering influence and informed decision-making.

THE LANGUAGE OF RESULTS: COMMUNICATING WITH EXECUTIVES

Executives prioritize decisions based on comprehensive business outcomes rather than isolated tactical data points. They focus on growth, cost management, risk mitigation, and value creation. Consider each of these four outcomes as a narrative framework. An effective business narrative features a clear protagonist (the initiative), a central challenge (the business problem), and a resolution (the measurable outcome). Executives instinctively follow these plotlines when evaluating proposals.

- **Growth:** How effectively are we accelerating customer acquisition, securing larger deals, penetrating new markets, or enhancing customer lifetime value? For a Chief Revenue Officer (CRO) or Chief Marketing Officer (CMO), this entails moving beyond activity reports (e.g., calls made, leads generated) to demonstrating a direct correlation between efforts and conversion rates, customer retention, and share of wallet. For a Chief Legal Officer (CLO), this means illustrating how onboarding or leadership development programs expedite employee productivity and revenue generation.
- **Cost:** How are we minimizing waste, improving operational efficiency, and reducing turnover or operating expenses? Chief Operating Officers (COOs) and Chief Financial Officers (CFOs) require evidence that efficiency improvements translate into higher margins or release capital for reinvestment. Learning leaders, IT directors, and supply chain managers need to demonstrate how their initiatives reduce rework, shorten cycle times, or prevent costly errors.

- **Risk:** How are we preventing compliance breaches, reputational damage, talent shortages, or operational disruptions? Risk extends beyond legal exposure to encompass everyday vulnerabilities that can jeopardize business stability. A marketing leader who compromises brand trust, a product leader who neglects quality control, or a CLO who treats compliance as a mere formality all create risks that can have far-reaching consequences across the organization.

- **Value:** How are we creating lasting benefits that extend beyond short-term financial results? Value represents the synthesis of growth, cost management, and risk mitigation, but it also encompasses the differentiated advantages that foster customer loyalty, employee engagement, and stakeholder investment. For a CEO or board of directors, this involves assessing whether initiatives strengthen brand equity, enhance organizational culture, and cultivate capabilities that are difficult for competitors to replicate. For a CLO, this involves demonstrating how learning programs build leadership pipelines, accelerate innovation, and prepare the workforce for future challenges, thereby transforming training from a cost center into a driver of enterprise value. For a Chief Information Officer (CIO), this involves showcasing how technology platforms enable scalability, data-driven decision-making, and digital trust that extend beyond simple cost savings. In today's matrixed organizations, value is amplified when collaboration transcends functional and regional boundaries. The ability to navigate the matrix and align diverse leaders on shared outcomes is what converts isolated successes into enterprise-wide advantages.

Every compelling results story follows the same consistent structure that Marcus used:

1. **Define the business problem:** Clearly articulate the stakes in terms of growth, cost, risk, or value.
2. **Describe the intervention:** Explain the actions your team took to address the problem.
3. **Highlight the consequence:** Show how the intervention impacted revenue, cost, risk, or value.
4. **Quantify and illustrate the business payoff achieved.**

When communicating updates, frame data as a *strategic* narrative rather than a mere data dump. Executives evaluate information through the lens of key organizational objectives: growth, cost reduction, risk mitigation, and value creation. If the relevance of your data to these objectives is unclear, you risk losing their attention. Therefore, translating data into a compelling narrative that elucidates who benefited, by how much, and why it matters is crucial for effective communication and lasting impact.

Growth

Executive reports invariably prioritize growth, focusing on top-line revenue, market expansion, and customer acquisition. Executives are keenly interested in whether the company is outperforming competitors and whether new strategies will capture market share. When presenting a growth-oriented narrative, begin by highlighting the potential risks to growth if no action were taken. Every compelling narrative requires stakes. Quantify the gap and then illustrate how your initiative effectively closed it. Executives are particularly receptive to narratives that demonstrate the transition from lost opportunity to captured value.

Simply stating, "We trained 200 account managers," fails to resonate with executives. They are more interested in knowing whether this training resulted in increased deal closures, expanded revenue per customer, or the capture of new opportunities.

At Hilti, reframing a sales training program to focus on contract conversions significantly changed the conversation. Instead of reporting, "We delivered training," we reported, "We increased conversion rates by 12%, resulting in $40 million in additional revenue in the first year." Consequently, executives shifted from questioning the budget to inquiring about the program's scalability.

The same principle applies to marketing. A CMO who boasts about generating 50,000 new leads will likely face scrutiny from the CEO regarding the actual conversion rate of those leads into opportunities and their impact on customer lifetime value. A pipeline filled with names is inconsequential unless it translates into tangible revenue.

Similarly, in product development, a CTO may highlight the number of features released in a quarter. However, unless those features drove increased adoption, expanded market reach, or strengthened customer loyalty, the roadmap reflects activity rather than strategic growth.

A client in global logistics initially measured success by counting the number of supervisors trained. However, reframing the data in business terms transformed the narrative: on-time delivery improved by 8% in regions where leaders received training, securing multimillion-dollar renewal contracts with two of their largest clients.

The 2024 PwC Global CEO Survey revealed that 83% of CEOs prioritize revenue growth among their top three objectives, even over cost reduction and risk management. Deloitte's 2025 Human Capital Trends study corroborated this, indicating that companies that explicitly link workforce or operational initiatives to growth strategies are 2.5 times more likely to surpass their peers in revenue expansion.

Translation in Action

Functional metric:

- 80 sales managers trained in 2 months
- 900 sales account managers completed digital training refresher module

Business translation:

- "$X above the baseline in new contracts closed after the sales manager training"
- "5,000 opportunities converted, producing $Y in pipeline growth from the 50,000 leads generated with this campaign"
- "Customer adoption rose by 12%, defending market share against our largest competitor"

Reflection

Which of your current initiatives could be reframed in terms of revenue growth, customer acquisition, or market share?

Cost

If growth is the first obsession of executives, cost is the constant pressure. Members of the C-suite live in the world of margins, operating expenses, and efficiency. They want to know: How does this initiative reduce spend? How does it make us leaner without sacrificing capability? The cost story is not about frugality. Instead, it is about resource liberation. Frame it as reclaiming capital or time that can be reinvested. The protagonist in a cost narrative isn't the expense cut; it's the capability you preserved while spending less.

At a manufacturing client, the learning team proudly reported that 10,000 employees had completed a new safety course. The CFO shrugged. When we reframed the metric, the story changed: "In the six months following training, workplace incidents dropped by 23%, resulting in $3.2M in savings in direct costs and another $1.7M in indirect costs due to the reduction in lost-time injuries." Suddenly, the CFO cared. It was not about training hours; it was about reducing costs that directly impacted margin.

At one organization that I worked with, we linked a frontline supervisor program to rework rates. Instead of saying, "We trained

450 supervisors," we said, "Rework decreased by 11%, which equated to $5M saved in material and labor costs." The program transitioned from being "another training initiative" to being a cost-control lever.

This lesson applies outside L&D. A COO might highlight utilization rates or contract labor spend, but true cost advantage comes when operations become more effective. While less expensive is good, maintaining the desired quality and timing is equally important. Similarly, a CMO who touts lower campaign spend without showing a reduced cost per acquisition has not proven real impact.

McKinsey's 2025 Cost Transformation Report showed that 77% of executives rank cost reduction as a top three priority during market downturns. Yet, fewer than 20% believed that their HR or operations teams could quantify their cost impact. Deloitte found that companies that connected workforce initiatives to efficiency gains were 2.3 times more likely to sustain profitability during volatile periods.

Translation in Action

To translate your activity measures to cost language, consider the following cash flow and efficiency lens:

Functional metric:

- Percentage reduction in processing time or cost per transaction.

Business translation:

- Working capital released, margin improvement, and liquidity preserved for reinvestment.

- A finance leader who automates invoice approvals can demonstrate a $2.4 million increase in available cash flow by reducing the cycle by 12 days. Rather than focusing solely on process efficiency, this leader should emphasize the capital freed up to fund innovation or mitigate market volatility.

Reflection

If your CFO requested an annualized cost savings report for the XX initiative, could you confidently provide one?

Risk

Risk is a primary concern for executives, encompassing financial risk, compliance failures, reputational damage, cybersecurity breaches, and talent shortages. The most compelling risk narratives precisely detail "near miss" scenarios. What potential negative outcomes were avoided due to your initiative? Executives engage when they can visualize the averted crisis and the resulting financial or reputational savings. How does this initiative safeguard the organization from disruption and prevent negative publicity?

In one instance, the L&D team at a financial services client initially presented compliance training as a routine requirement. Although participation rates were high, policy violations persisted. The CLO reframed the narrative: "This program reduced policy violations by 30%, decreasing the company's potential regulatory exposure by $18 million." This shift moved the focus from training completion rates to tangible risk mitigation. Instead of simply encouraging employees to share answers or quickly complete the module, managers began emphasizing the severe consequences of policy violations, underscoring the critical need for behavioral change.

During the Hilti Evolution rollout, the primary risk was operational rather than regulatory. Failure to adopt the new opportunity-to-close process within the sales team would have resulted in the loss of millions in pipeline visibility. By establishing local advocates and implementing a globally supported learning framework, the risk of adoption failure was successfully mitigated.

This pattern extends across various functions. CIOs often report system uptime metrics, but executives are more interested in how improved reliability reduces the risk of customer churn. CMOs may track brand mentions, but the board also prioritizes the avoidance of reputational risks through strategic crisis communication. COOs might present supply chain throughput data, but the CEO requires assurance that disruptions in one region will not cripple the entire operation.

PwC's 2025 CEO Survey revealed that 49% of CEOs anticipate a "major disruption" within the next five years, primarily citing cyber risk, regulatory change, or talent shortages. However, less than 25% expressed confidence in their strategies to mitigate these risks. Korn Ferry's 2025 Global Talent Risk Report demonstrated that companies that link workforce initiatives to risk management experienced 30% fewer compliance incidents and 22% lower turnover in critical roles.

Translation in Action

To connect activity metrics to risk mitigation, consider the following example:

Compliance, Safety, IT, or Legal Functional Metric:

- Number of incidents avoided, policies updated, or audits passed.

Business Translation:

- Reduced risk exposure, stabilized insurance premiums, and prevented downtime.
- A compliance officer who tracks "zero late filings" is not merely managing paperwork; they are safeguarding millions in potential fines and preserving brand trust. Reframe your narrative by asking: What risk did we mitigate from the balance sheet this quarter? This

approach transforms your metrics from sounding like prevention to reflecting performance.

Reflection

Which risks are most relevant to your company (regulatory, operational, reputational, or talent) and how do you demonstrate their mitigation?

Value

Ultimately, public companies exist to create value for shareholders, and private companies answer to owners or boards who measure long-term return. Your data should illustrate how each initiative increases enterprise value and strengthens confidence in your strategic direction.

A value-driven narrative connects the current quarter to long-term objectives, linking today's metrics to future advantages and articulating why the company will remain relevant, trusted, and profitable in the years ahead. The most compelling storytellers in this context function less as reporters and more as architects of enduring success.

In one company I worked with, the CLO reframed leadership development as "building the leadership pipeline that will secure our succession plan." This resonated with the board, signaling stability and reassuring investors of the company's ability to sustain performance.

At Hilti, our accelerated adoption of the Nuron battery platform was not solely about training the sales team on new products. It was also about protecting market share in a highly competitive industry and increasing long-term enterprise value by differentiating through innovation. By demonstrating that L&D's contribution expedited time-to-market and preserved customer loyalty, we directly linked learning to shareholder confidence.

The same principle applies across departments. A CMO's brand campaign is significant when it elevates Net Promoter Scores and boosts market valuation. A COO's supply chain redesign is impactful when it signals resilience to investors. A CIO's tech platform is valuable when it underpins scalability and digital transformation, driving enterprise growth.

A 2024 Harvard Business Review report indicated that companies with strong leadership pipelines and succession planning were 3.5 times more likely to maintain stock price stability during market shifts. Accenture's 2025 report on Human Capital and Enterprise Value found that talent-related initiatives now account for nearly 70% of total enterprise value in knowledge-intensive industries. In essence, the way you develop and deploy talent is directly correlated with shareholder value.

Translation in Action

Value can be the most challenging concept to convey. Use this example to consider how to reframe your data into value-creation terms.

Functional metric:

- Number of compliance trainings completed or audits passed.

Business translation:

- Reduction in legal exposure, enhanced market credibility, and secured license to operate.
- When a quality team maintains ISO certification, they are ensuring the company can bid for contracts worth tens of millions. The metric is not "certified auditors trained" but "revenue streams protected."

Reflection

If your board asked, "How are we securing the talent, systems, and capabilities that will drive shareholder confidence?" what evidence would you present?

Growth, cost management, risk mitigation, and value creation constitute the four key narratives for executive audiences. When data informs a compelling narrative, the focus shifts from justifying work to cultivating belief. Belief, in turn, is the true currency of executive decision-making.

TRANSLATION, NOT REINVENTION

By now, the underlying pattern should be evident. Every executive-level discussion revolves around a narrative of tension, translation, and consequence. The objective is not to rewrite data, but to reframe the story it conveys. Articulating results does not necessitate abandoning functional metrics. Completion rates, engagement surveys, and uptime percentages remain relevant as inputs or proof points that substantiate the narrative, rather than serving as the narrative's focal point. When translating data, adopt a declarative and confident tone: "This improved X by Y, resulting in Z." Avoid hedging or uncertainty. Results-oriented narratives establish credibility through clarity, not modesty.

The critical skill lies in translation. An effective leader can correlate functional outcomes with the priorities of executive leadership. Consider these examples from the field of Learning & Development:

- A 10% improvement in onboarding completion is not merely about training efficiency; it represents accelerated time to productivity, which, in turn, generates revenue more rapidly.
- A 15% increase in engagement scores is not simply about "happier employees"; it signifies reduced turnover, which

translates to millions in savings in recruiting and
retraining expenses.

- A 20% improvement in system uptime is not merely an IT
  achievement; it signifies fewer customer disruptions,
  which enhances retention and loyalty.

Each function possesses its unique translation narrative. The
examples below illustrate how the same narrative arc from activity to
outcome to enterprise value manifests across various contexts:

Sales/CRO: An increase in demo volume does not, in itself, indicate sales success. The key metric is the conversion rate of those
demos: a four-point improvement in win rates, accelerated pipeline
velocity, and a favorable margin mix indicative of deals closing with
enhanced profitability. When Sales correlates activity metrics with
profitable conversion, sustainable pipeline velocity, and margin
growth, it evolves from a volume-driven engine to a driver of enterprise value.

Marketing/CMO: An increase in Marketing Qualified Leads
(MQLs) may appear impressive on a dashboard, but if those leads
stagnate, they are inconsequential. The salient narrative emerges
when sales-accepted opportunities increase, customer attrition
declines, and the pipeline-to-revenue conversion rate improves.
When Marketing correlates campaign metrics with customer retention, brand trust, and revenue expansion, it transitions from a cost
center to a catalyst for market growth.

Product: Releasing more features is not, in itself, indicative of
innovation. Success is measured by adoption rates, increased
customer retention, reduced churn, and growth in annual recurring
revenue (ARR) as customers derive lasting value. When Product
correlates feature delivery with adoption, customer stickiness, and
recurring revenue, it evolves from a release-driven operation to a
catalyst for competitive advantage.

COO/Operations: High utilization rates can be misleading, as
they may be achieved by overextending personnel without corresponding gains in efficiency, ultimately harming the business. The

focus should be on increased throughput with on-time, in-full delivery, sustained quality, and improved contribution margins. When Operations effectively links efficiency gains to throughput, quality, and contribution margin improvement, it transitions from a mere production engine to a critical pillar of profitability.

Supply Chain: While reporting on-time delivery rates may suffice for operational dashboards, the true measure of success lies in improved perfect order rates, reduced reliance on costly expedited freight, and higher contract renewal rates driven by customer trust in reliability. When Supply Chain connects fulfillment accuracy to customer reliability, renewal rates, and cost avoidance, it evolves from a logistics function to a vital guardian of business continuity.

Customer Success: Simply counting customer touchpoints or positive feedback does not demonstrate tangible impact. Results are realized when these interactions translate into higher gross retention, stronger net revenue retention (NRR), and measurable expansion of revenue per account. When Customer Success connects service activity to retention, net revenue growth, and customer lifetime value, it transcends the role of a relationship manager, becoming a steward of enterprise resilience.

Human Resources: Tracking training completions, headcount, or time-to-hire may appear efficient on paper, but these are inputs, not outcomes. The true value is revealed when talent initiatives reduce regrettable attrition, succession pipelines shorten the time to fill critical roles, and leadership bench strength facilitates faster execution of strategic priorities. When HR connects people metrics to revenue protection, growth capacity, and risk reduction, it moves beyond a support function to become a key enabler of enterprise performance.

Finance: Reporting solely on budget variance or expense reduction provides an incomplete picture. True financial leadership is demonstrated when intelligent capital allocation fuels innovation, working capital efficiency frees up cash for growth, and predictive analytics proactively prevent margin erosion. When Finance directly links cost discipline and forecasting accuracy to enterprise agility and shareholder confidence, it transitions from a gatekeeper to a strategic growth partner.

Executives do not expect you to reinvent your function; they expect you to translate its impact into their language.

When you communicate in the language of results, three key outcomes occur:

1. **Budgets shift from cost centers to strategic investments.** Executives prioritize funding initiatives that deliver outcomes aligned with their strategic objectives.
2. **Influence expands across the organization.** Stakeholders engage you earlier in the process, recognizing your ability to connect functional expertise to broader strategic goals.
3. **Legacy is built as a driver of the business, not just a function.** You evolve from simply managing a function to actively driving overall business success.

Therefore, communicating in the language of results is not merely a communication technique; it is a fundamental leadership responsibility. Without it, even the most well-designed programs or initiatives risk marginalization. With it, your work becomes indispensable. Translation transforms information into influence, and when executives incorporate your phrasing into their own updates, it signifies that your message has become ingrained in the organization's understanding of its business.

PRACTICAL FRAMEWORK: REFRAMING

There are several steps to reframing in a way that adds value to your narrative.

Begin with the CEO's perspective

To gain executive buy-in, align your communication with their established priorities. CEOs consistently reiterate their top three to five objectives, embedding them within the organizational culture. These priorities, often referred to as pillars, north stars, or imperatives, ultimately address the core question: "What must we achieve this year to succeed?"

Functional teams often overlook this crucial alignment, focusing on internal projects rather than broader organizational goals. Executives invariably assess presentations by asking: "Does this facilitate the faster, more efficient, or less risky achievement of our top priorities?" While functional leaders possess in-depth knowledge of their areas, they may overestimate the audience's familiarity with their specific terminology.

Consider your metrics as a foreign film. The executive audience is engaged, but comprehension requires subtitles in their language, which are the CEO's priorities. Without this context, the message appears disjointed. Framing initiatives within the CEO's priorities transcends mere alignment; it constructs a narrative. It establishes the context (the CEO's strategic ambition), introduces the subject (the initiative), and demonstrates the resolution (how the team accelerates the realization of that ambition).

A learning and development leader I coached had prepared a proposal for enhanced leadership training. The presentation was polished, and the program was well-designed. However, it lacked a direct connection to the CEO's primary objective: entering two new markets within 18 months. By revising the opening, we shifted the focus from "We need to build more leadership skills" to "We need leaders equipped to execute rapidly across functions to achieve our new-market goals." Same program, reframed message. The CEO immediately engaged, and approval followed.

The 2024 PwC CEO Survey revealed that over 80% of CEOs concentrate their efforts on a select few key priorities. Yet, a 2025 RedThread Research report indicated that only approximately 20% of support function leaders consistently align their initiatives with these priorities. This discrepancy explains why leadership often perceives many functions as reactive rather than strategic.

Practical Application

1. Compile the last two to three CEO addresses or quarterly updates.
2. Upload these documents to an AI-powered analysis tool.
3. Prompt the tool to: "Analyze these files and identify the most frequently repeated keywords to define the language for my next presentation."
4. Practice articulating your initiatives using these same keywords.

Instead of: "We trained 400 new managers on feedback skills."

Say: "We prepared 400 new managers to lead high-performing teams, directly supporting our CEO's second priority: customer retention."

Strategic Reflection

If your CEO reviewed your latest project presentation, would they immediately recognize its connection to one of their stated top priorities?

Translate Functional Metrics into Business Metrics

Each function typically has its own scorecard. HR tracks engagement, IT monitors uptime, L&D measures completions, and Procurement analyzes savings. While these metrics hold significance within each function, they often lack resonance in the boardroom. If Sales were to present leads without linking them to revenue growth, the reception would be similar. This is because these metrics are not expressed in the language executives use for decision-making.

Imagine delivering a presentation in English to an audience in Paris that speaks only French. While the content may be accurate, it will not be understood. Functional metrics are like English in this scenario, while business metrics serve as the French subtitles that executives require. When translating metrics, focus on narrating the journey, rather than simply presenting the data point. For instance, "We reduced downtime" can be transformed into "We recovered six days of customer availability, saving $2 million in lost orders." This conveys the same information but with a different narrative arc, one that executives are more likely to remember.

At a global consumer goods company, the HR leader proudly presented an "85% engagement score." The CEO paused and inquired, "So what? Did that reduce our attrition problem in Asia? Did it improve productivity in our plants?" The HR leader had not prepared the translation, and the moment passed, negatively impacting her credibility.

In contrast, an IT leader at a tech firm effectively translated his metrics. Instead of reporting "99.9% uptime," he informed the CEO, "We reduced outages that used to cost $2 million annually in lost transactions, putting that money back into the company's pocket." The data was the same, but one presentation used functional terms, while the other focused on business results.

Deloitte's 2025 Human Capital Trends study revealed that only 16% of executives believe HR and L&D leaders "effectively translate their metrics into business impact." Conversely, organizations where support functions routinely do this are 2.4 times more likely to be rated as "strategic partners" by their boards.

Practical Application

1. Begin with your functional metric.
2. Ask: "If this improved by 10%, what would it mean for growth, cost, risk, or shareholder value?"
3. Convert it into the outcome. Example:

- "90% completion rate" → "Onboarding time cut by 20%, accelerating revenue by $X."

- "Employee engagement +8 points" → "First-year attrition down 15%, saving $Y in hiring costs."

Strategic Reflection

Review your last three reports. How many metrics would be readily understood by your top stakeholders without translation?

Quantify, Even Imperfectly

For decades, L&D has struggled to prove causation. Leaders debate whether a sales increase resulted from training, new marketing campaigns, or improved products. The ROI Institute and Kirkpatrick models have influenced this thinking, urging practitioners to isolate impact. However, in reality, business results rarely stem from a single factor. Growth, retention, and productivity are the result of multiple initiatives working in concert.

Executives understand this. They do not need you to prove that your function alone caused the result. They want to see that you contributed to it, that the trend moved in a positive direction, and that your initiative correlated with that improvement.

At Tucanes, the training manager declined to present beyond completion rates, citing an inability to demonstrate causation. This missed an opportunity to highlight potential correlations with improved sales cycle times. In contrast, the Head of Training at Vida LLC acknowledged the multifaceted drivers of turnover reduction, including compensation adjustments, cultural initiatives, and training. Instead of pursuing elusive causation, they positioned training as one component of a comprehensive strategy that reduced attrition by 18%, resulting in millions of dollars in savings for the company. This framing proved effective due to its realism and transparency.

While precision is valuable, persuasion is paramount. Executives appreciate confidence tempered with honesty, rather than unattainable statistical perfection. They seek credible connections, not irrefutable proof, understanding that business decisions often rely on correlation.

Practical Application

- For L&D, shift the focus from isolating training's direct causal impact. Following a thorough root cause analysis, demonstrate how your initiatives correlate with improvements in key business areas: growth, cost optimization, risk mitigation, or value creation.
- Across all functions, contextualize your metrics with relevant business data, such as attrition trends, regional revenue figures, and productivity scores.
- Communicate with clarity: "While multiple factors contributed, our function's initiatives aligned with and supported this positive trend."
- Utilize established models to structure your analysis, but avoid allowing them to become a hindrance. Executives favor transparent correlations over silence justified by false precision.

Strategic Reflection

Identify instances where you've hesitated due to an inability to prove direct causation. Consider alternative approaches to showcase correlations that enhance, rather than diminish, your credibility.

Cultivating the Translation Habit

Isolated translation efforts are insufficient to alter perceptions. To reshape executive views of your function, consistent and ongoing translation of activities into tangible business results is essential.

When coaching my team, we initiated a practice of delivering concise, three-minute pitches. Each project lead was tasked with articulating how their initiative connected to a strategic business priority, using the organization's strategic language as a central reference point. Initially, these sessions were marked by awkward pauses, with team members defaulting to metrics like "training hours" or "engagement scores." This "rehearsal stage" is a necessary component. By practicing the narrative aloud, identifying jargon that diminishes impact, and refining the business case, language evolves from mere translation to an intuitive skill. Consider it akin to developing the muscle memory required for effective business storytelling. Over time, team members presented to executives with confidence and fluency.

This process extends beyond refining presentations; it's about reshaping identity. When your team consistently communicates in terms of business outcomes, executives transition from viewing you as order-takers to trusting you as strategic partners. Your team members, in turn, begin to see themselves as drivers of business performance, proactively adapting their roles to contribute to organizational objectives.

Practical Application

- **Role-play:** Conduct role-playing exercises simulating leadership conversations. Task your team with pitching projects to the CFO or CEO.
- **Integrate Translation into Project Approval Processes:** Ensure that no initiative progresses without a clear, concise statement articulating its connection to growth, cost management, risk mitigation, or shareholder value creation.
- **Recognize Effective Translation:** Publicly acknowledge and celebrate team members who demonstrate excellence in reframing metrics to align with business outcomes during meetings and other relevant forums.

- **Reinforce Translation Through Repetition:** Consistently challenge teams to articulate the business impact of their work by repeatedly asking, "So what? What is the business value?" until these connections become second nature. Employ methodologies such as the "5 Whys" to facilitate a deeper understanding of root causes and ensure a focus on fundamental outcomes.

Strategic Reflection

Assess the frequency with which you and your team translate your work into business-oriented results. Establish a consistent cadence to cultivate this practice as a standard operating procedure, rather than an ad-hoc exercise before executive reviews. Translation is realized when your team members shift their focus from reporting functional activities to narrating the broader business story. When this occurs, metrics evolve from simply quantifying effort to articulating tangible value.

LEADING THROUGH DOWNTURNS AND SHOCKS

In times of economic downturn or unforeseen crises, a results-oriented approach becomes paramount. When revenue declines, markets fluctuate, or disruptions impact business operations, executives face a common challenge: dwindling resources, heightened scrutiny, and a prioritization of survival. Downturns serve as a critical test of your strategic narrative. During prosperous periods, mere activity can be mistaken for genuine progress. However, in challenging times, only tangible outcomes matter. The narrative that endures is the one that demonstrates how your efforts mitigate risk and facilitate recovery.

In such circumstances, leaders who focus solely on activity metrics tend to fade into the background. Executives are less interested in the number of training hours delivered, demos scheduled, or new features released. Instead, they seek insights into investments that safeguard cash flow, maintain customer loyalty, and create opportunities for recovery.

This was evident during the 2008 financial crisis, when many organizations abruptly reduced learning budgets. Chief Learning Officers (CLOs) who had justified their programs based on training completion rates saw entire initiatives eliminated. Conversely, those who had linked their work to increased revenue, reduced employee turnover, expedited onboarding processes, or enhanced customer retention not only weathered the storm but, in some cases, even secured additional funding. Their leadership recognized that compromising these capabilities would impede recovery efforts. This pattern recurred during the 2012 downturn and again during the COVID-19 shutdowns, when organizations were compelled to fundamentally alter their operating models.

This pattern extends across various functional areas. During crises, executives' focus narrows, prioritizing stability, cash flow, and continuity. Consequently, your communication style must adapt accordingly. Employ concise language, quantify impacts in monetary terms, and emphasize verbs of protection ("preserve," "retain," "sustain") to establish credibility during periods of heightened emotional stress.

- **Sales:** In downturns, executives do not reward mere activity. They inquire, "Which accounts are profitable, which are eroding margins, and how are we safeguarding customer lifetime value?"
- **Marketing:** Vanity metrics, such as impressions and clicks, lose relevance under the pressure of budgetary constraints. What endures are campaigns directly correlated with pipeline conversion and customer retention.
- **Product:** A roadmap brimming with features is inconsequential if adoption remains stagnant. Leaders must demonstrate how each release bolsters market share

or generates efficiencies that customers are willing to pay for.

- **Operations and Supply Chain:** Utilization statistics or throughput rates are of limited value if quality deteriorates or costs escalate. The focus shifts to reliability, flexibility, and contract renewals based on customer trust, even amid turbulence.

Economic downturns expose vulnerabilities and compel leaders to demonstrate their ability to drive resilience. Those who have consistently translated their impact into tangible outcomes related to growth, cost management, risk mitigation, and value creation become indispensable. Leaders who fail to adapt face marginalization, budget cuts, and diminished team morale. Consequently, they struggle to regain traction during the recovery phase, having been perceived as detached from the core business operations. The key takeaway is that articulating results is not merely a fair-weather practice; it serves as an insurance policy that safeguards your function during times of adversity. It transforms your role from expendable to essential.

By now, it should be clear that communicating results is a prerequisite for executive-level engagement. When you can effectively link your work to growth, cost, risk, and shareholder value, executives take notice. Polite acknowledgment gives way to genuine interest, and you are invited to contribute to critical decisions that shape the company's trajectory. In the aftermath of a crisis, leaders remember those who communicated with clarity and composure, using sound business principles, while others succumbed to panic. This fosters trust, which, in turn, leads to inclusion. The narrative you construct during a downturn often determines your presence at the table when growth returns.

During my time at Hilti, I witnessed this firsthand when John Willox assumed leadership of Global L&D. With a background in Sales and Marketing, including his previous role as Marketing Manager in one of our most dynamic business units, John had cultivated a strong network of allies throughout the organization. He possessed a deep understanding of Hilti's customers and effectively integrated this perspective into his leadership of L&D, fostering a thriving environment. We consistently prioritized the customer in every Executive Board presentation, despite our focus on employee training. This customer-centric approach enabled us to make business-relevant decisions, frame our narratives around the ultimate impact of our training programs, and gain access to nearly all executive meetings. L&D was no longer relegated to the periphery. During the revision of our fundamental sales program, the heads of Hilti's two largest regions actively participated by role-playing the content in front of their colleagues. This transformation in perception from an afterthought to active involvement underscores the power of effective communication. John established a solid foundation upon which we could build, leading to the team's ongoing and future success.

John also demonstrated overcoming the limitations of individual achievement. Even with mastery of results-oriented communication, influence is restricted when the message is delivered in isolation. Genuine transformation necessitates allies. When leaders from finance, operations, sales, HR, and IT begin to echo your narrative, reinforcing the business impact of your work in their own dialogues, your influence is amplified. In our situation, attempting to independently implement the new sales training program within a company that considered its B2B sales approach a key differentiator would have been unsuccessful. Securing the endorsement of individuals who had risen through the ranks to assume key leadership roles and who also recognized its value and advocated for it proved invaluable.

∼

LEVERAGING AI AS A STRATEGIC THOUGHT PARTNER

Employ AI to gain insights into the metrics that resonate most with executive leadership by analyzing their existing communications. Similar to how an effective coach helps clarify one's message, AI can serve as a "storytelling mirror," reflecting the language used by executives. This enables you to evaluate whether your narrative aligns with their perspective or remains confined to functional jargon.

1. **Compile Executive Communications:** Collect transcripts or recordings of recent presentations by key stakeholders, such as C-suite executives, including quarterly earnings calls, all-hands meetings, and board updates. Input these transcripts into an AI platform.

2. **Extract Key Executive Metrics:** Prompt the AI to: "Identify the key metrics [Executive Name] emphasized in this transcript and categorize them under growth, cost, risk, and value." This will reveal the areas of focus for leadership. As you review the output, pay close attention to the tone and sentiment expressed, in addition to the specific terminology used. Note whether executives convey a sense of urgency, optimism, or caution. Mirroring this emotional register in subsequent presentations will foster subconscious alignment more effectively than simply matching metrics.

3. **Overlay Team Metrics:** Upload your team's data and prompt the AI to: "Identify areas of overlap or disconnect between our current metrics and these executive-level metrics." This step will highlight whether your team is tracking data that is relevant to executive leadership or expending resources on inconsequential information. When presenting these findings, maintain a human-centric approach. Instead of attributing the insights solely to AI, frame them as "Our review indicated ..." AI should function as a behind-the-scenes analyst, with your judgment and voice taking center stage in the narrative.

4. **Incorporate Cross-Functional Data:** Expand your analysis beyond your team's reports. For example, when evaluating training data, solicit relevant metrics from other departments:

- **From Sales:** "Did leads generated after sales training demonstrate a higher conversion rate to opportunities?"
- **From HR:** "Did the time-to-fill for open positions decrease following the implementation of interview skills training?"
- **From Finance:** "Did supervisor training result in a reduction in overtime costs or rework?"

5. **Analyze Granular Metrics:** Go beyond high-level metrics and examine more detailed indicators. These are the metrics that executives are already closely monitoring.

By leveraging AI in this manner, you are not prompting it to illuminate existing alignments and discrepancies within your work. In this context, AI serves as a tool to refine intuition, enabling you to assess whether your narrative resonates with the metrics and sentiments deemed critical by key decision-makers. This is what distinguishes AI as a genuine thought partner, transcending the capabilities of a mere spreadsheet.

Reflection

If you compared your current functional metrics with the data presented in the last three stakeholder presentations, what degree of overlap would be evident? Where might you identify potential misalignments in your measurement approach?

The following chapter explores how to cultivate a network of allies who champion your message in your absence.

6

CULTIVATING ALLIES TO MULTIPLY YOUR INFLUENCE

E stablishing a network of allies will amplify your voice and unlock opportunities. Credibility is not earned in isolation; it is earned within a community, when others attest to your impact based on their observations of how it drives their success. These allies facilitate the achievement of your objectives and ensure that your team members recognize the intrinsic value of networking. The objective is not merely to expand your contact list, but to comprehensively understand the organization, make informed decisions, and drive tangible results. This skill is mission-critical for enabling functions such as HR, Finance, IT, and Legal, among others, whose results are often imperceptible to the P&L. Articulating results-oriented communication is how you render your function comprehensible to the business.

Chapter 6 will guide you in how to:

- Identify the most critical allies.
- Develop partnerships that transcend transactional support.
- Create a coalition of advocates who ensure your function remains central.

Communicating in the language of results earns you a seat at the table. Cultivating allies ensures you maintain it.

I once collaborated with a CLO who possessed exceptional strategic acumen. She could correlate any learning initiative to business outcomes and was fluent in results-oriented communication. She excelled at connecting disparate elements to create innovative solutions. However, her proposals rarely gained traction at the executive level. The reason? She was the sole advocate. She lacked both allies and a framework to enable others to perceive how it could positively impact their objectives.

Eivind Slaaen, a former colleague at Hilti, employed a different strategy. Before proposing a program, he consulted with key stakeholders across the organization, engaging the CFO on cost implications, the COO on operational risks, and the CHRO on talent pipelines. Consequently, by the time his proposal reached the executive level, these leaders were already articulating its merits in their own terms, integrating their unique perspectives. The proposal resonated as a collective vision rather than an individual initiative. While some team members occasionally felt that outcomes were being compromised, the reality was that he was ensuring broader participation in shaping those outcomes.

Again, the differentiating factor was not competence, but rather, the cultivation of allies. Previous chapters emphasized the importance of reframing language, such as shifting from "training" to "transformation" and from "activity" to "results," to establish credibility. However, credibility alone is insufficient. Organizational influence is amplified through strategic networks.

When cross-functional leaders champion a message and underscore how a function contributes to their success, influence is exponentially increased. The focus shifts from justifying value to having others advocate on your behalf.

However, not all initiatives gain traction. I recall a promising leader, Angelica Houston, who assumed responsibility for DEI training. Driven by personal experiences of discrimination, she was deeply committed to driving meaningful change. Angelica promptly engaged external vendors specializing in female equality, delivering well-researched and heartfelt programs. However, she made a critical error: she operated in isolation.

She did not engage Finance or HR to explore how DEI could mitigate costly turnover, nor did she consult with Operations on how inclusion could foster psychological safety, leading to improved outputs and innovation. Sales was not involved to explore how diverse teams could better connect with customers. While she solicited input from perceived underrepresented groups, she failed to integrate their feedback into a cohesive, actionable strategy.

The outcome was predictable. In a company that was 80% male, many felt their concerns were excluded. People of color felt their experiences were overlooked, while LGBTQ employees felt pressured to conceal their identities. Men felt their perspectives were disregarded. Across the workforce, managers dismissed the training as inconsequential compliance exercises disconnected from business objectives.

Employees passively completed the modules and attended the sessions to comply with regulations or avoid negative publicity, but the initiative failed to resonate. Without allies across the organization to shape, refine, and champion the program, Angelica's passion became a liability. Instead of driving transformation, the program reinforced the perception of DEI as a mere HR requirement rather than a critical business imperative.

The lesson is clear: even the most vital initiatives can be marginalized without allies. Passion without partnership undermines credibility rather than building it.

Without allies, your influence is tenuous. Network science has demonstrated that impactful leaders are defined less by their titles and more by the breadth and diversity of their connections. Research on organizational networks indicates that high-performing leaders cultivate relationships across at least three functions outside their own, and these connections account for up to 40% of their effectiveness. Similarly, a 2024 McKinsey study on enterprise agility concluded that companies with robust cross-functional networks were twice as likely to outperform peers in decision-making speed and innovation. In essence, the breadth of your allies directly correlates to the breadth of your influence.

Early in my leadership career, I focused on delivering exceptional training programs and articulating their value, underestimating the importance of building relationships across the organization. I resisted investing time in networking. Fortunately, strong mentors helped me realize that credibility hinges as much on the opinions of others when you are absent as it does on your presentations.

At Hilti, with some initial reluctance and encouragement, I made it a practice to proactively engage with leaders across various departments. I inquired about the challenges faced by sales directors, even when they weren't seeking support. I asked operations leaders about inefficiencies, and finance leaders about their financial concerns. During each site visit, I made it a point to connect with new individuals at every level. I scheduled virtual meetings with people who could broaden my understanding of the organization, even if there was little direct overlap with Learning & Development. By demonstrating this approach, I encouraged my team to dedicate time to networking, expanding both their business acumen and their professional network.

My objective in these meetings was to understand their perspectives. Over time, these conversations fostered trust. Consequently, when I proposed a learning initiative, it was perceived not merely as an L&D solution, but as a collaborative solution to a shared problem. They supported it, championed it, and reinforced its importance in their own discussions with executives.

IDENTIFYING THE RIGHT ALLIES

Not all relationships are equally valuable. Some colleagues, while supportive, may have a limited view of the company that can constrain your perspective. Others may be skeptics whose endorsement could significantly shift the conversation.

Begin by mapping your organization's power dynamics. Identify those who are accountable for the CEO's key objectives, who have the ear of the Board, and whose input carries significant weight in decision-making. Conduct this exercise collaboratively with your team members to gather diverse perspectives and equip them with the skills to navigate organizational networks effectively.

During my tenure at Hilti, I observed that sales directors wielded considerable influence, possessing unparalleled field expertise and the unwavering trust of the General Manager across departments. Securing their advocacy proved as vital as crafting a well-structured business case, a dynamic mirrored in the finance department. The CFO's endorsement of an initiative held significantly more weight than my individual advocacy.

Therefore, the initial step in cultivating alliances lies not in indiscriminate networking but in pinpointing those key leaders whose support can champion your work in arenas beyond your direct reach.

RECIPROCAL RELATIONSHIPS ARE ESSENTIAL

Frequently, functional leaders approach potential allies with a unilateral agenda, seeking assistance solely for program approval, goal attainment, or data provision. This approach can swiftly erode credibility.

Genuine alliances flourish when reciprocity is present. When engaging with operations leaders, I prioritized understanding their challenges before advocating for training programs. By inquiring about project bottlenecks and areas of inefficiency, I fostered an environment of trust. Often, these discussions revealed solutions unrelated to training, such as process improvements, cultural adjustments, or leveraging existing training resources. Consequently, when a training solution proved relevant, they were more receptive, recognizing my commitment to their success.

A similar approach was adopted with finance and marketing.

Rather than solely seeking cost data, I offered assistance in enhancing the capabilities of their analysts and strengthening their teams. By collaborating on report analysis and data origins, I demonstrated that I valued their expertise as key partners, rather than expecting them to merely support my objectives. Building alliances requires a commitment to giving before asking.

MAINTAINING YOUR NETWORK

Cultivating alliances is an ongoing process, requiring continuous investment. To prevent relationships from deteriorating, I scheduled quarterly conversations with at least three leaders outside my immediate function, focusing solely on understanding their evolving priorities and challenges.

These conversations served as an invaluable early-warning system, providing insights into business shifts before they surfaced in broader meetings. I gained advance notice of impending changes and identified activities perceived as effective. This proactive alignment of learning initiatives ensured that my allies were already advocating for my proposals, as they resonated with their priorities.

Research supports this approach. Deloitte's 2023 Human Capital Trends report indicates that informal networks, such as impromptu conversations, brief check-ins, and cross-functional collaborations, are stronger predictors of organizational resilience than formal reporting structures. Similarly, CEB (now Gartner) research demonstrates that high-performing managers dedicate up to 30% more time than their peers to cultivating cross-functional relationships, a time investment directly correlated with accelerated problem-solving and improved business outcomes. Networks serve to disseminate not only information but also credibility.

Early in my career, I mistakenly equated self-reliance with strength, perceiving requests for assistance from other functions as admissions of weakness. When I first involved Finance in an L&D project, I anticipated criticism. Instead, their insights revealed opportunities, fundamentally changing my perception of partnership. True influence begins when you relinquish the impulse to protect your domain and embrace collaboration.

PRACTICAL FRAMEWORK: CULTIVATING ALLIES TO AMPLIFY YOUR INFLUENCE

We form and join many communities in various areas of our lives. We usually do this intentionally to expand our connection within our community, to support our children's school system, to reinforce our sense of belonging, or to follow our passions. In this chapter, we are only discussing the network that you build to amplify your leadership legacy and influence and how you model that to your team members.

Map Your Organization's Influence Network

While formal organizational charts exist, the true map of influence often differs significantly. The key question is not merely hierarchical reporting lines but rather who shapes critical decisions. For example, ABB used a Power/Interest grid that categorized stakeholders by their level of influence (power) and their level of concern (interest) and then designated engagement levels for each depending on their designation.

At Hilti, I quickly recognized the sales directors and CFO as central figures in the influence network. Their endorsement could either ensure the success or doom the failure of any proposal. For certain projects, such as sales, their concern was high. For other projects, such as leadership development, their concern would be moderate.

Practical Application

- Outline your organization's "influence network." Identify the top five individuals whose opinions your CEO values

most.
- Determine the points of intersection between your function and their priorities.
- Prioritize these areas. Influence spreads most effectively when initiated at the source.

Reflection

Within your organization, who could express a single sentence about your function in a meeting that would carry more weight than your entire presentation?

Prioritize Inquiry

Leaders are adept at discerning whether your motivation is to promote an agenda or to genuinely seek understanding. Remember that the primary purpose of networking is to enhance your comprehension of the organization, broaden your perspective, and strengthen your strategic insight. Allies are not simply won over; they are cultivated through a deep understanding of their situation and the development of solutions tailored to their specific needs. Listening is the foundation upon which alliances are built.

I recall a conversation with an operations leader who expressed frustration with repetitive project delays. He came in requesting project management training. Instead of immediately agreeing, I inquired, "What is slowing you down?" He detailed bottlenecks, resource allocation challenges, and a lack of cross-training. He explained how these delays got worse when pressure increased. Only then did I suggest a potential solution. By prioritizing curiosity over a pre-determined agenda, I established myself as a trusted partner.

Practical Application

- In your initial three meetings with potential allies, commit to asking more questions than you answer.
- Utilize prompts such as: "What is your biggest challenge this quarter?" or "What obstacles are hindering your progress toward your goals?"
- Take thorough notes and maintain an open mind. Your objective is not to solve their problems directly but to offer fresh perspectives and help them identify potential solutions they may have overlooked.

Reflection

When was the last time you met with a peer leader with the primary intention of learning?

Give Before You Ask

The most effective way to build trust is to provide value before seeking assistance. At Hilti, I initiated workshops on presentation skills for marketing analysts. While not directly aligned with the learning agenda, this initiative addressed a need within the marketing team and benefited their leadership. This resulted in increased productivity and accelerated product-related decisions, mitigating previous rework cycles. Consequently, when I later required their support for a learning management system (LMS) pilot program, they readily collaborated.

Practical Application

- Identify a specific area where your team can provide value to another department, regardless of the scale.

- Share relevant insights, offer valuable resources, or provide a capability that streamlines their work processes.

Reflection

What is one tangible way you can contribute to a peer's work this month without expecting anything in return?

Maintain Consistent Connection

Building alliances cannot be achieved during times of crisis. As previously mentioned, I prioritized scheduling quarterly conversations with leaders outside my direct function, without a predetermined agenda. I also encouraged my team members to adopt this practice when visiting sites or seeking to expand their professional network. While these conversations did not always yield immediate urgent matters, they frequently revealed underlying issues that could be addressed proactively.

Practical Application

- Establish a regular cadence: aim for one networking conversation per month with individuals outside your direct function, unrelated to pending requests.
- Keep it concise: allocate 30 minutes focusing on understanding "What's changing in your world?".
- Document key insights and identify patterns that align with your function's impact.

Reflection

Which three leaders have you not engaged with in the past six months, but would benefit from doing so?

Transform Allies into Advocates

An ally is someone who offers support. An advocate is someone who champions your cause when you are not present. The transition from ally to advocate occurs when leaders recognize their own success is linked to yours. At Hilti, sales directors became advocates when they realized that training initiatives accelerated their progress toward achieving revenue targets. They went beyond merely supporting my programs; they actively promoted them.

Practical Application

- After establishing trust, invite allies to share their perspectives in executive meetings.
- Provide them with a concise, impactful statement they can use: "This initiative will reduce onboarding time by four weeks and accelerate revenue generation."
- Recognize and reinforce their advocacy by publicly acknowledging their contributions.

Reflection

Who already recognizes the value of your function, and how can you empower them to effectively communicate that value to others?

ADVOCATES GENERATE SUSTAINABLE INFLUENCE

Leadership research has extensively documented the transition from support to sponsorship. Catalyst's longitudinal studies on sponsorship reveal that individuals with active advocates are 23% more likely to be promoted, as sponsors leverage their resources to cultivate opportunities. Similarly, Korn Ferry has observed that organizations where leaders actively champion cross-functional colleagues experience a 34% increase in success rates for major transformation initiatives. Furthermore, a Harvard Business Review analysis of advocacy networks indicates that projects with cross-functional champions are twice as likely to progress beyond the pilot stage.

In essence, passive support from allies is insufficient; active advocates who publicly connect your work to tangible business results are essential for establishing lasting influence. Cultivating allies transcends mere office politics; it is a matter of scalability. No leader, regardless of their expertise, can unilaterally reshape organizational perception. When allies across manufacturing, marketing, sales, finance, operations, and HR echo your message and champion your initiatives as their own, your influence is amplified. You transition from a solitary voice to a collective chorus. These advocates will also be instrumental in developing well-rounded talent throughout the organization, providing valuable insights during talent review sessions, and identifying prospective talent. This extends beyond individual projects or initiatives, ensuring the effective development of the next generation of leaders by modeling behaviors aligned with broader organizational objectives.

~

LEVERAGING AI AS A STRATEGIC THOUGHT PARTNER

While allies are forged through relationships, AI can facilitate the identification of strategic investment areas. The natural inclination to gravitate towards individuals similar to ourselves or those we find personally agreeable can create blind spots in perceiving the individuals who can contribute to a more comprehensive understanding of the organization. Instead of relying on guesswork to identify influential figures, you can leverage existing company data to pinpoint where support is most impactful.

To effectively utilize your AI thought partner, consider the following steps:

1. **Mine Executive Communications:** Upload transcripts of recent presentations from key stakeholders and pose the question: "Which leaders, departments, or initiatives are most frequently mentioned in connection with company priorities?" This will reveal existing areas of formal influence.

2. **Analyze Cross-functional Documents:** Input strategy documents, project updates, or meeting summaries, and inquire: "Which individuals or teams are most frequently referenced as critical to success?" This process can uncover hidden influencers who may lack formal titles but significantly impact decisions.

3. **Overlay Your Initiatives:** Provide the AI with concise descriptions of your current projects and ask: "Which executives' priorities align most closely with these initiatives?" This will highlight individuals who are already invested in outcomes that your function supports.

4. **Simulate Ally Conversations:** Once potential allies are identified, prompt the AI: "Assuming you are [ally's name], what three questions would you ask me before supporting this project?" Use these insights to refine your approach and frame your initiatives in a manner that resonates with their priorities.

5. **Identify Potential Blind Spots.** Consider: "Which executives or functions appear underrepresented in these initiatives, and what potential risks could this create?" This helps ensure a comprehensive and inclusive network.

Reflection

If you were to apply this process, which key stakeholders would emerge, and how does this align with your existing network?

Research validates this approach. A 2024 MIT Sloan study on digital collaboration revealed that AI-driven analysis of communication patterns can identify "hidden influencers" whose impact on decision-making surpasses their formal position. Similarly, RedThread Research has demonstrated how AI, when applied to talent and project data, can pinpoint "connectors" who effectively bridge silos and drive change. By leveraging AI, organizations can move beyond assumptions and gain data-driven insights into the true dynamics of influence.

7

CULTIVATING A LASTING CULTURE

I once worked with a senior executive who possessed strong business acumen, financial expertise, and operational discipline. However, his team operated in an environment of fear. Meetings were fraught with tension, communication was guarded, and innovative ideas were suppressed. While his performance metrics appeared favorable, his leadership style stifled innovation.

In contrast, I worked with another leader who fostered a completely different environment. He was not only respected but also trusted. His team understood his expectations for results, but they also knew he valued their growth and input. This balance of accountability and empathy cultivated a culture of willing and capable contribution. Employees embraced risk, shared new ideas, and the business flourished.

The differentiating factor was not strategy, as both leaders operated from a similar framework. Rather, it was the culture they cultivated. Culture extends the impact of decisions beyond the meeting room, shaping whether employees view requests or processes as bureaucratic hurdles or shared responsibilities. Allies secure representation, while advocates amplify the message. Influence extends beyond individual relationships. Enduring leaders cultivate a culture where their core values are embedded into the organization's daily operations.

Early in my career, I witnessed the impact of culture firsthand when I walked into a manufacturing plant a week after a layoff. The usual sounds of laughter were replaced by an atmosphere of fear and uncertainty. As a young manager, I lacked the experience to address the employees' concerns and focused solely on maintaining productivity. With experience, I now recognize the importance of bringing people together, addressing the situation transparently, and acknowledging what I did and did not know.

While allies can provide short-term support and advocates can drive initiatives forward, embedding a vision into the culture is essential for creating a lasting legacy that extends beyond one's tenure and sustains the organization through change. In this chapter, we will explore how leaders translate strategy into tangible behaviors, reinforce values through organizational systems, and cultivate a culture that serves as a competitive advantage rather than a superficial facade. Allies amplify influence; culture multiplies legacy.

CULTURE AS A STRATEGIC LEVER

Culture is often informally defined as "how we do things around here." However, this description understates its significance and can be weaponized against innovation and diverse perspectives. Culture is the operating system of an organization, dictating how strategy is executed, how individuals behave in the absence of oversight, and how resilient the organization is when facing disruption.

When leaders deliberately shape culture, they extend their influence beyond individual departments, programs, or initiatives. They establish conditions where desired behaviors become the norm, not the exception. When culture reinforces strategy, the organization accelerates progress, adapts rapidly, and sustains performance.

Conversely, an unattended culture can become a liability, fostering toxicity, driving away talent, and stifling innovation. Leaders who achieve short-term targets can inadvertently leave the organization weaker than they found it.

Culture is not solely an HR concern; it is the responsibility of every executive. While seemingly intangible, data demonstrates that culture is a strong predictor of performance. McKinsey's 2024 State of Organizations report found that companies with strong, adaptive cultures were 3.7 times more likely to report top-quartile performance than those without. Cultural health was identified as the single most significant factor linked to organizational resilience.

Stories in Action: Culture by Design

When NovaEdge, a retail technology company, faced its most significant disruption in a decade, its leadership team initially faltered. A new competitor had entered the market with a product that was more affordable, user-friendly, and rapidly gaining customer loyalty. The CFO advocated for cost reduction and increased efficiency, while the COO pushed for accelerated production of the existing product line. The CMO argued for increased marketing expenditure. Each executive clung to familiar strategies. Except for Maya Patel, the Chief Transformation Officer.

Maya had established a reputation for her insatiable curiosity. She had consistently encouraged her teams to challenge assumptions, experiment with novel approaches, and, most importantly, discard practices that were no longer effective. Her guiding principle was simple: "What got us here won't get us there."

Instead of defending outdated strategies, Maya initiated a "reset lab" with her colleagues. For three days, senior leaders from across the company set aside their dashboards and status reports to examine live customer feedback, competitor demonstrations, and AI-driven analyses of internal performance data to identify emerging patterns. The lab's ground rules were clear: no defending the past, only exploring the future.

Initially, the atmosphere in the room was tense, with some executives resisting change. "We cannot discard our established processes," they argued, "they define our identity." Maya, however, adopted a different approach. She posed the question: "If we were starting anew today, what practices would we immediately discontinue?" Gradually, others began to participate. The CMO conceded that increasing marketing expenditure without corresponding product improvements would only exacerbate customer dissatisfaction. The COO recognized that accelerating production without redesigning the product would merely amplify its existing weaknesses. The CFO, upon analyzing the correlation between customer attrition and outdated features, acknowledged that efficiency could not compensate for irrelevance.

By the conclusion of the reset lab, the team had eliminated three long-standing processes, expedited two experimental designs, and reorganized cross-functional teams to facilitate the testing and implementation of improvements within a 90-day timeframe. While challenging, the initiative proved effective. Within six months, NovaEdge successfully reversed customer attrition and regained market share, exceeding all expectations.

The pivotal factor was not the reset lab itself, but rather the culture that Maya had cultivated in advance of the crisis. She had nurtured a culture where leaders were expected to rethink, relearn, and reshape, and where questioning the status quo was viewed not as disloyalty, but as essential for survival. Culture transcends being a mere backdrop; it encompasses the stage, the script, and the spotlight. Optimal strategies, products, or training initiatives are unsustainable without a supportive culture.

- A culture that rewards compliance fosters minimal effort.
- A culture that rewards curiosity fosters innovation.
- A culture that rewards advocacy fosters multipliers.
- A culture that rewards accountability with empathy
  fosters growth and retention.

Leaders possess the opportunity to shape culture, whether consciously or not. Leadership is defined as much by what is tolerated as by what is celebrated. *The choice lies between shaping culture by default, allowing habits and power dynamics to dictate the tone, or shaping it by design, aligning culture with strategy and values.*

The success of the NovaEdge reset lab was not attributable to skillful facilitation, but to a pre-existing culture of readiness. Maya had dedicated years to fostering an environment where individuals could challenge established assumptions without fear, where intelligent failure was expected, and where performance was measured by both results and adaptability.

This is the critical lesson that leaders often overlook: strategy and governance are important, but not self-sufficient. Culture acts as the multiplier, determining whether governance becomes a perfunctory exercise or a genuine source of alignment, and whether the organization executes strategy with vigor or disregards it as a fleeting trend.

At Hilti, I witnessed this firsthand. The company's culture of care and performance fostered a sense of accountability. Employees expected high standards, but also knew that leadership was invested in their development and well-being. This equilibrium fostered a willingness to push boundaries, experiment with new concepts, and accept feedback. Governance was effective because it was supported by the culture.

In contrast, organizations where culture undermines strategy often face setbacks. For instance, a DEI program, though initiated with good intentions, may falter if it lacks inclusivity across the broader workforce. Similarly, a compliance initiative focused solely on training, without engaging managers in consistent reinforcement, will likely stall. In both scenarios, the underlying issue is not the content itself but the prevailing culture.

Culture is critical because it shapes the context in which decisions are made. An elegantly designed governance process will be ineffective if individuals do not feel secure enough to voice their opinions, leading challengers to remain silent. Likewise, a well-crafted strategy will suffer if employees do not believe their contributions are valued by leadership, resulting in a lack of enthusiasm in execution.

PRACTICAL FRAMEWORK: HOW LEADERS SHAPE CULTURE

Culture is shaped less by what leaders say and more by how they behave under pressure. The most effective leaders do not choose between performance and people; they integrate both. Accountability without empathy breeds fear, while empathy without accountability erodes trust and results. Sustainable performance emerges when leaders hold high standards and demonstrate genuine care in equal measure.

Model Accountability and Empathy

During my time at Hilti, I observed that the most effective leaders were those who balanced accountability with empathy. They challenged their teams to achieve ambitious goals while also providing support, listening to their concerns, and acknowledging their lives outside of work.

One sales director, in particular, consistently drove significant growth year after year. Simultaneously, he demonstrated genuine care for his team members. When someone faced difficulties, he responded not by lowering expectations but by investing in coaching and support. This balance fostered loyalty, and his team's exceptional performance stemmed from a desire to succeed, not merely an obligation. Consequently, that region consistently outperformed others.

Google's Project Aristotle also highlighted this dynamic. Initially, the assumption was that the success of their highest-performing teams was attributable to the right combination of technical expertise and exceptional team leadership. However, Project Aristotle revealed that psychological safety (a culture where team members feel comfortable speaking openly, admitting mistakes, and challenging ideas without fear) was the primary factor. This cultural foundation amplified the impact of all other elements. Teams characterized by psychological safety consistently surpassed their peers in innovation, problem-solving, and overall performance, underscoring the importance of culture in maximizing the value of technical talent.

Korn Ferry's 2025 Global Talent Report further substantiates these findings, revealing that organizations where leaders actively balance accountability with care experience 22% higher retention rates among high-potential employees and outperform their peers by a factor of 2.3 in profitability. Employee retention yields direct cost savings by mitigating turnover expenses, including:

- **Recruitment and hiring:** The Society for Human Resource Management (SHRM) estimates that the cost of replacing an employee can range from 50% to 200% of their annual salary. For a highly skilled or executive employee earning $100,000 annually, replacement costs could range from $50,000 to $200,000.
- **Training and onboarding:** The process of integrating new hires, encompassing training programs and the productivity deficit during the learning curve, constitutes a significant expense.
- **Lost institutional knowledge:** The departure of a high-potential employee results in the loss of valuable knowledge related to company processes, clients, and organizational history, which can impede operations and negatively impact client relationships. Culture, therefore, serves as a competitive differentiator.

Beyond mere cost avoidance, retaining high-potential employees enhances profitability through multiple avenues:

- **Increased Productivity:** Engaged and experienced employees demonstrate higher productivity. Research from Gallup indicates that companies with highly engaged employees experience a 21% increase in productivity and a 22% rise in profitability. High-potential employees, by definition, deliver more value than their average counterparts, with some sources suggesting they can provide up to 91% more value.

- **Business Growth:** A stable and experienced workforce facilitates more effective strategic planning and consistent project delivery. This, in turn, drives revenue growth and contributes to long-term stability.
- **Enhanced Valuation:** Companies with strong employee retention often command higher valuations from investors. One study revealed that companies with robust retention practices received valuations of 6–7 times their EBITDA, compared to 2–3 times EBITDA for those with high turnover. For a company with $3 million in EBITDA, this discrepancy can result in a valuation gap of $12 to $15 million.

Culture is a critical business imperative.

Practical Application

1. Identify a specific behavior that, if more prevalent within your organization, would enhance one or more characteristics of psychological safety, thereby improving the balance of accountability and care.
2. Consistently model this behavior in your own work and during meetings.
3. Draw attention to your actions to emphasize their importance.
4. Recognize and acknowledge instances of this behavior occurring throughout the organization.

Reflection

Which behavior could you personally model more effectively to help align the culture with your desired state?

Embed Values in Daily Behaviors

Culture is not defined by posters or values statements. It is shaped by the behaviors that leaders reward, tolerate, or ignore. Ignoring a behavior is often perceived as tolerating it, leading to its increased prevalence. Leadership is defined by both the behaviors you celebrate and those you tolerate.

During the Evolution rollout, rather than mandating the new process and measuring compliance, the global team reinforced values by empowering local advocates. This daily behavior of trusting local leaders and granting them ownership signaled that collaboration and respect were not mere slogans, but genuine principles.

Conversely, the case of Wells Fargo illustrates how a flawed culture can undermine governance, resulting in a toxic environment. Despite having processes, targets, and compliance checks in place, Wells Fargo's culture incentivized cross-selling at any cost. Employees who failed to meet sales targets were shamed, penalized, or replaced, while those who achieved targets were celebrated, even if their practices were questionable. This resulted in millions of fraudulent accounts, billions in fines, and lasting reputational damage. This was not due to a lack of rules or governance, but rather a toxic culture that amplified undesirable behaviors.

Practical Application

1. Translate your values into 2–3 observable behaviors.
2. Reinforce these behaviors in meetings, recognition programs, and performance reviews.
3. Hold yourself accountable: ignoring behaviors that contradict your stated values shapes the culture by default.

Reflection

Which behaviors within your team most clearly reflect your values, and which undermine them?

Build a Culture of Advocacy

Upon Satya Nadella's appointment as CEO, Microsoft was burdened by internal competition and siloed operations. The prevailing culture, which emphasized individual brilliance over collaboration, stifled innovation. Nadella redefined the corporate culture around a growth mindset, fostering an environment that valued continuous learning over static expertise. Leaders began to model vulnerability, curiosity, and empathy, which facilitated more rapid innovation and collaboration across product lines. These cultural shifts unlocked latent potential within the technical teams, resulting in enhanced effectiveness. Consequently, Microsoft's market value more than tripled within a few years, demonstrating the transformative power of culture in overcoming stagnation.

At Hilti, similar dynamics were observed. When individuals felt integral to a collaborative effort focused on a tangible objective, and when they believed that recognition for success would be shared, they became strong advocates for the program or initiative. This sense of ownership was further amplified when advocates from within their own ranks modeled and promoted the change. For example, the endorsement of training programs by sales directors led to a significant increase in adoption rates, highlighting the impact of internal advocacy over top-down mandates.

Practical Application

1. Identify influential advocates within each major function or location.

2. Provide them with compelling narratives and resources to articulate the importance of the initiative.
3. Publicly acknowledge and celebrate their advocacy to reinforce its value.

Reflection

Who are the natural advocates within your organization, and how can you empower them to amplify your message?

Leverage Culture as an Early-Warning System

Initially, quarterly meetings with leaders from outside my direct functional area were not conceived as culture-building exercises. However, I later realized that these interactions significantly shaped the organizational culture. By actively listening to the changes and challenges within their respective domains, I signaled the importance of cross-functional collaboration. Addressing their concerns proactively fostered a culture of responsiveness and cooperation. These meetings provided early micro-signals about successes and failures, often before formal metrics were available. This allowed for a continuous assessment of organizational health, the identification of potential issues, and the proactive shaping of narratives.

Practical Application

1. Establish regular check-ins with leaders outside of your immediate function.
2. Inquire about their shifting priorities and emerging trends.
3. Communicate how your work will adapt to support their needs, thereby demonstrating culture in action.

Reflection

How frequently do you receive insights into business shifts before they are presented in executive summaries?

Balance Global Alignment with Local Ownership

One of Hilti's core strengths was its ability to harmonize global strategic objectives with localized execution. The successful rollouts of Evolution and Nuron were predicated on the understanding that a uniform approach was not universally effective. Local leaders were provided with the necessary tools and frameworks but were entrusted to adapt the delivery to suit their specific markets. By providing key messages and content while encouraging localization, we effectively balanced the need for speed and global alignment with the nuances of local terminology and expectations.

This approach is practiced across many industries. For example, McDonald's and Starbucks incorporate locally relevant menu items across the globe. One notable example is Unilever's Connected4-Growth approach, exemplified by Dove's "Real Beauty" campaign. This strategy fosters a sense of familiarity while acknowledging local cultural nuances. Observing this balance internationally evokes a sense of comfort derived from both the familiar and the novel.

This equilibrium between global alignment and local ownership cultivates a culture of consistency and respect, encouraging participation through inclusivity.

Practical Application

1. Establish non-negotiable standards at the global or organizational level.
2. Empower local or functional leaders to adapt the practical implementation of these priorities.

3. Implement feedback loops to ensure local adaptations inform the broader organizational system.

Reflection

Assess areas where control may be excessive and areas where greater alignment is needed.

Treat Culture Like a System, Not a Slogan

Many companies reduce culture to a superficial marketing campaign. However, true culture resides within organizational systems, including hiring practices, promotion criteria, performance reviews, and recognition programs. If a company espouses innovation but penalizes risk-taking in performance evaluations, the system will inevitably override the stated values. Systems are dynamic and require adaptation; consequently, organizational culture must also evolve.

In its early stages, Uber's culture was defined by relentless hustle and an unrestricted drive for growth. This approach yielded rapid global expansion and market dominance. However, it also fueled scandals related to harassment, legal disputes, and leadership crises, eroding the trust of investors and regulators. Upon assuming the role of CEO, Dara Khosrowshahi initiated a cultural reset, integrating accountability, ethics, and employee voice alongside performance metrics. While Uber continued to prioritize growth, its culture now framed it as sustainable and responsible. Without this cultural reset, the company's governance and strategy would have likely collapsed. Culture facilitated its recovery.

Practical Application

1. Conduct a comprehensive audit of existing systems to

determine whether they reinforce or contradict the desired culture.

2. Adjust hiring, promotion, and reward practices to align with the desired culture.

3. Engage employees in co-creating cultural norms to foster ownership and commitment.

Reflection

Identify organizational systems that currently strengthen the culture and those that undermine it.

CULTURE IS INTEGRAL TO LEADERSHIP

Culture is not merely a byproduct of leadership; it is an integral aspect of leadership itself. Leaders who intentionally shape culture extend their influence beyond immediate financial results. They cultivate organizations that attract talent, retain customers, and thrive amidst disruption. Conversely, when leaders neglect culture, toxicity proliferates, turnover accelerates, and even the most effective strategies falter.

A leader's legacy will be defined by both the results achieved and also the culture left behind, as that culture determines the long-term sustainability of those results.

∽

LEVERAGING AI AS A STRATEGIC LEVER

Culture is not a declaration; it is a lived experience. As a leader, you shape it daily, intentionally or unintentionally. AI can serve as a mirror, not by prescribing a culture, but by assessing whether the culture you aspire to create aligns with the reality your people experience.

1. **Compare Words with Actions:** Upload your company's values statement or strategic plan, alongside transcripts from recent all-hands meetings, CEO town halls, or departmental updates. Then, query the AI: "What are the most prevalent themes in leadership communication, and how well do they align with our stated values?" This will reveal whether your organizational systems reinforce or contradict your declared culture.

2. **Identify Functional Discrepancies:** Compile concise summaries from sales, operations, and HR detailing their most significant achievements and challenges this quarter. Input these summaries, along with your cultural priorities, into the AI. Ask: "Where do these narratives reinforce our intended culture, and where do they expose inconsistencies?" This will highlight areas where the culture is robust and areas where it is faltering.

3. **Evaluate Behavioral Consistency:** Provide the AI with a brief overview of a recent decision you made as a leader. Then, pose the question: "Given that our culture emphasizes accountability and empathy, how does this decision either reinforce or contradict those values?" This does not replace personal reflection but enhances it.

4. **Uncover Blind Spots:** Ask the AI: "Which groups or perspectives appear underrepresented in our cultural narrative?" This often identifies overlooked functions, geographies, or demographics whose cultural experiences diverge from the dominant majority.

Reflection

If you were to compare your company's stated values with the daily experiences of your employees, what discrepancies would emerge, and what role could you play in bridging those gaps?

∾

A LEADERSHIP CULTURE CHECKLIST

Culture transcends posters, slogans, and speeches; it is embedded in the daily operational mechanisms of your business. As a leader, every decision regarding hiring, promotions, recognition, or tolerated behaviors either reinforces or undermines the culture. Utilize this checklist to evaluate whether your mechanisms align with the culture you aim to cultivate.

Hiring Standards	
Do our interview questions assess values as rigorously as skills?	
Are we clear on the behaviors that are considered non-negotiable?	
Do we have someone designated to veto a candidate who does not align with our culture?	
Reinforcement: Culture begins with the individuals you welcome into the organization.	
Promotion Criteria	
Are promotions based on a combination of results and the methods used to achieve them?	
Do we reward leaders who demonstrate a balance of accountability and empathy?	
Do we remove individuals who achieve targets but compromise the culture?	
Reinforcement: Culture is defined by what you celebrate and what you tolerate.	
Meeting Structure	
Do our meetings exemplify the culture we aspire to create?	
Are meetings utilized to clarify priorities or simply to report activities?	
Do we provide a platform for dissenting opinions, or only for those who agree?	
Reinforcement: Culture is evident in how decisions are made and whose voices are valued.	
Post-Mortems & Reviews	
When projects fail, do we learn collaboratively as a team?	
Do we examine behaviors and assumptions that led to outcomes?	
Do we implement lessons learned in subsequent cycles?	
Reinforcement: A culture of accountability and growth thrives on reflection, not scapegoating.	
Recognition Systems	
Who is recognized most often: individual contributors or collaborative teams?	
Do we recognize behaviors that align with our values, not just results?	
Is recognition public, consistent, and linked to the behaviors we want to encourage?	
Reinforcement: Recognition is a powerful tool for signaling cultural values.	

Reflection Questions

- Which mechanisms within my team are reinforcing the desired culture and which are silently undermining it?
- What is one mechanism I can adjust in the next 30 days to better align culture with strategy?

8

WHEN TECHNOLOGY AMPLIFIES YOUR LEADERSHIP

C ulture is the organizational heartbeat that shapes behavior and workflow. However, even a strong culture requires supportive systems to maximize its impact. Technology is a critical system that leaders can leverage to accelerate results. In today's environment, successful leaders view technology as an integral part of the leadership ecosystem.

This chapter explores technology as a multiplier, examining how leaders can evaluate tools, frame investments, and integrate systems to enhance capacity, uncover insights, and reinforce strategy. Technology implemented without purpose creates noise, while technology applied with discipline and intent amplifies leadership.

LEADING WITH DATA

The leadership team of a mid-sized manufacturing company convened for their quarterly review. They took pride in their culture of collaborative, customer-focused, and resilient teamwork. However, the mood shifted when the CFO presented the financial results, which revealed eroding margins, increased customer churn in a key market, and rising onboarding costs for new hires.

The CEO turned to the CHRO and CLO, stating, "Despite our investment in our strategy, I am not seeing the anticipated returns. What are we missing?" One leader responded, presenting a narrative centered on technology, not as a novel system, but as a leadership enabler.

She explained how her team had piloted an AI-enabled platform to assess onboarding effectiveness. This initiative provided concrete data, demonstrating that new hires reached full productivity in eight weeks instead of twelve, resulting in millions in early revenue capture. First-year attrition decreased by 15%, and supervisors reported a reduction in plant safety incidents.

She then synthesized her observations. "This extends beyond the software itself. It's about leadership efficacy. Technology has provided a level of visibility previously unattainable, affording us the opportunity to refocus on strategic initiatives. Moreover, it has enabled us to reinforce our espoused values of accountability and care. Previously, we operated on conjecture. Now, we lead with informed insight."

The room's attention sharpened. For once, the discourse on technology transcended the conventional concerns of cost, complexity, or compliance, and centered on the essence of leadership.

Realizing Potential Through Technology

My initial realization of the extent of untapped data occurred during a straightforward interview. Following the development of a Training Management System (TMS) for a client, we sought to validate our assumptions. I posed a direct question: "What is the time and cost associated with onboarding training administration?"

One client possessed the answer. They had meticulously tracked hours, calculated expenses, and could provide a precise quarterly cost for onboarding administration. However, this client proved to be an exception. In subsequent interviews, leaders conceded their lack of awareness. Onboarding administration was not tracked; it was tacitly integrated into existing roles. The actual cost remained obscured within job descriptions. More surprisingly, few possessed data on program effectiveness. Programs were routinely modified or adjusted, but without a baseline for comparison, decisions were made without empirical support.

This is where technology proved transformative. By assisting these organizations in establishing a baseline for tracking onboarding duration, cost, and subsequent new hire performance, we generated previously unseen data. We then guided them in systematically modifying variables, inputting performance data into AI. Over time, clear patterns emerged, delineating which changes enhanced performance, which were neutral, and which were wasteful.

The objective was not to supplant personnel with technology, but to liberate them from guesswork. Instead of relying on intuition or emulating competitors, leaders could discern what strategies were effective within their specific context. They transitioned from debating opinions to making evidence-based decisions.

The transformation was profound. Executives who had previously dismissed onboarding as a mere HR detail began to recognize its potential as a lever for productivity and retention. Leaders who had regarded training administration as inconsequential background activity began to view it as an area ripe for savings and impactful change.

The core message is clear: technology does not replace the role of individuals in onboarding; it amplifies it. By revealing hidden costs and demonstrating tangible performance patterns, technology elevates onboarding from an invisible function to a strategic discussion at the executive level.

Highlighting Blind Spots Through Technology

The onboarding scenario highlighted a common issue: many organizations operate with critical blind spots. Despite significant investments in training, systems, and processes, leaders often fail to track the true costs and measure the effectiveness of these initiatives. Consequently, decisions are frequently based on instinct, opinion, or the influence of dominant voices, rather than data-driven insights.

While this approach may have been adequate in the past, when markets evolved at a slower pace, today's business environment demands agility. Strategy cycles are now measured in months, not years. Competitors emerge rapidly, and entire industries face disruption from technological advancements and evolving workforce dynamics. The relentless pace of change necessitates a more strategic approach.

In this context, technology, particularly AI, is not merely a luxury but a critical enabler of leadership impact. As the CLO in that quarterly meeting recognized, technology transcends dashboards and efficiency gains; it amplifies the effectiveness of leadership. Technology optimizes resource allocation. Manual processes that once required weeks of analysis can now be completed overnight, freeing up time for strategic initiatives. While this efficiency is often taken for granted, the next step involves leveraging technology to surface insights. By identifying patterns in data, leaders can gain a clearer understanding of what is working and what is not. This enables them to scale impact, reach a wider audience more quickly, and ensure the consistent delivery of their message and intent.

In one company I consulted, well-intentioned trainers made annual adjustments to the onboarding program, adding new modules, removing sessions, or changing facilitators. However, the impact of these changes on performance was never measured. Each iteration was essentially an experiment without supporting data. As a result, first-year employee turnover remained stubbornly high, and time-to-productivity lagged behind competitors.

The leadership team was competent but lacked the tools to gain visibility into performance drivers. In contrast, organizations where we established performance baselines and integrated data into AI systems experienced a transformative shift. Leaders could finally answer critical questions:

- What is the true cost of onboarding?

- What were the actual costs associated with developing that product?
- When should a salesperson be generating revenue in our industry?
- What level of scrap is acceptable in this manufacturing facility?
- Which elements enhance productivity, and which are time-consuming?
- How do new hires who complete this version of onboarding perform compared to previous cohorts?

Instead of relying on subjective opinions, leaders began making decisions based on empirical evidence. This fundamental shift improved the quality and focus of executive discussions. Research supports the assertion that organizations using AI to support leadership decisions achieve greater speed and improved performance:

- Deloitte's 2025 Human Capital Trends report indicates that 76% of executives are using AI to inform decision-making, and those organizations are 2.5 times more likely to report significant productivity gains.
- MIT Sloan Management Review (2024) found that companies applying AI to leadership data experienced a 30% reduction in strategy execution cycles due to their ability to identify patterns and make adjustments proactively.

According to RedThread Research (2025), organizations leveraging AI for talent analytics are 3.4 times more likely to identify high-potential leaders early, directly impacting pipeline strength and retention. The message is clear: leaders who embrace AI are not relinquishing their judgment but augmenting it.

AI-DRIVEN LEADERSHIP INSIGHTS

AI transforms the landscape by processing vast datasets beyond human capacity and revealing patterns across organizational silos. This includes identifying correlations between sales performance and training programs, attrition rates and onboarding processes, or system uptime and customer loyalty, all of which are insights of paramount importance to executives. Critically, leaders who effectively partner with AI will maintain their relevance, while those who do not risk falling behind.

Many leaders are initially drawn to AI for its ability to automate tedious tasks, such as report generation, calendar management, and compliance reminders. While valuable, this is merely the starting point. When our Training Management System was developed, the immediate benefit was automation. The system eliminated manual participant management, trainer scheduling (and rescheduling), spreadsheet-based onboarding tracking, and the need to solicit updates from managers. This freed up considerable time for more strategic, value-added tasks.

However, many organizations stop here, treating AI as a mere replacement for clerical work and missing its true potential. The real power of AI lies in its ability to accelerate leadership thinking. Beyond automating reports, AI connects disparate data points, revealing correlations between onboarding and retention, training modules and sales conversion, and system uptime and customer loyalty.

In this way, AI serves as a force multiplier, enhancing time management, sharpening judgment, deepening insights, and enabling more strategic leadership conversations when used judiciously. This approach to AI has been demonstrated throughout this book, and building familiarity will enable you to further utilize tools specific to your function.

REAL-WORLD APPLICATIONS

I have always been fascinated by the diverse reactions people have to technology. When introducing a new tool, some exhibit curiosity, while others display resistance. Personally, I do not experience this fear and wonder why curiosity does not override it. Recognizing the power of the brain's fight-or-flight response, I have learned that to overcome resistance, it is essential to create experiences that reframe technology as a source of safety and opportunity rather than a threat.

For me, technology has always represented predictability and progress. It eliminates subjective interpretations and biases; when errors occur, they can be easily corrected. This consistency is reassuring and allows us to design work that leverages human strengths, such as creative thinking, judgment, and empathy, while entrusting repetitive tasks to technology. When leaders embrace innovation in this way, they empower individuals by automating them into positions of greater possibility, rather than automating them out of relevance.

For decades, I have recognized technology's transformative potential. Early in my career, I transitioned from printing installation manuals to distributing them via CD-ROM, resulting in significant savings of paper, printing time, and packaging resources. This shift improved efficiency and demonstrated a respect for time and environmental stewardship. I remain excited by technological advancements that enhance clarity, speed, and the human experience in our work.

Before embracing new technologies, I consider the potential impact on my team and address these key questions:

- What anxieties or challenges might team members face during this digital transition?
- How can curiosity, predictability, and incremental achievements highlight the opportunities presented by these changes?

From Video Production to Strategic Influence

Arpita Banerjee, an instructional designer I collaborated with, exemplified this approach. Instead of relying on traditional video production timelines, she mastered AI-powered tools such as Synthesia, WellSaid, Canva, and Adobe Premiere. This enabled her to rapidly transform even short excerpts from Teams calls into professional, engaging training videos. By leveraging AI, she maintained consistent character sets across video series, integrated diverse concepts, and enhanced interactivity.

The result was a significant reduction in production time, from weeks to hours, and a marked increase in relevance. The ability to incorporate video snippets from subject-matter experts lent authenticity and immediacy to the content.

Because Arpita had demonstrated the value of her innovative approach, I readily approved her experimentation. This resulted in faster completion of training modules, reducing costs and improving timeliness and relevance. The accelerated production cycle ensured that training aligned with real-time needs.

This represents the shift from automation to acceleration: moving beyond merely performing existing tasks faster to enabling entirely new ways of working that enhance credibility and influence.

Executives prioritize contributions to growth, cost savings, and risk mitigation. While automation offers efficiency, which is now a baseline expectation, acceleration drives effectiveness, positioning your function as a strategic asset.

The most compelling evidence of technology's ability to amplify leadership comes from organizations that have successfully leveraged it to drive tangible results. In each case, success stemmed not from simply adopting new tools, but from strategically integrating them into the organization's culture and leadership priorities.

Public Examples

At IBM, leadership recognized onboarding as an opportunity to accelerate productivity and shape company culture. By leveraging Watson's AI to personalize the onboarding experience, they reduced time-to-proficiency by 40% and increased productivity by 35%. While the technology was a key enabler, the underlying success factor was leadership's strategic use of AI to reinforce the values of speed and personalization.

Hitachi encountered a similar challenge: new employees experienced protracted onboarding, while managers devoted substantial time to answering repetitive inquiries. By implementing an AI-enabled internal assistant, they reduced onboarding time by four days per new hire. This not only freed up HR capacity but also cultivated a culture of efficiency and support, enabling leaders to concentrate on higher-value strategic priorities.

In the financial sector, AIG, under the leadership of CEO Peter Zaffino, leveraged AI to enhance underwriting and operational processes. Generative models not only accelerated task completion but also provided leadership with improved predictability and deeper insights into risk management. The key to success was not simply the algorithms themselves, but rather how leadership employed them to bolster resilience and foster confidence throughout the organization.

Even consultancies are raising concerns about technology implementation. Deloitte's 2025 Human Capital Trends report advocates for a revised approach to evaluating technology investments. Leaders should shift their focus from "What's the latest tool?" to "How does this technology empower our people to perform better, faster, and more sustainably?" This perspective positions technology as a strategic leadership decision rather than merely a technical upgrade.

The common denominator in these examples is clear: technology alone does not guarantee results. Leaders who strategically align technology with their organization's strategy, culture, and governance transform it into a powerful multiplier. Those who fail to do so risk treating it as a mere system upgrade.

PRACTICAL FRAMEWORK: AI AS A MULTIPLIER

AI transcends its role as a mere time-saving tool. When strategically deployed by leaders, it amplifies their influence across five key dimensions, with each dimension building upon the previous one to drive impact.

Freeing Capacity

Technology can unlock latent capacity within your organization. In one client company, the Learning and Development (L&D) team dedicated approximately 40% of its time to manual reporting tasks, including compliance tracking, course completion monitoring, and spreadsheet consolidation. Following the implementation of an AI-enabled SaaS platform, these tasks were automated, freeing up significant time. What previously required a full week of work for two analysts now runs automatically overnight, allowing for the reinvestment of time into strategic initiatives.

This experience is not isolated. Deloitte's 2025 Human Capital Trends study revealed that organizations automating administrative HR and L&D tasks reallocated an average of 35% of staff time to strategic initiatives. In essence, automation streamlines operations and provides leaders with the bandwidth to focus on higher-value activities that drive business growth.

For example, one learning services organization automated tasks such as grading, time tracking, and report formatting, resulting in dozens of hours reclaimed each week. These automations served as the foundation for redirecting resources toward client-facing initiatives, innovation, and talent development, leading to increased employee satisfaction and improved profit margins.

Practical Application

1. Conduct an audit of your team's activities to determine the allocation of time between tactical and strategic tasks.
2. Identify 2–3 repetitive processes that are suitable for automation. Don't forget to just cut any processes that are unnecessary or no longer adding value.
3. Reinvest the time saved into projects that are directly aligned with business outcomes.

Reflection

If you could free up 20% of your team's time tomorrow, what strategic initiative would you prioritize?

Surfacing Insights

Technology offers the capability to rapidly identify insights that might otherwise be overlooked. In interviews regarding onboarding practices, the majority of leaders acknowledged the absence of a baseline for effectiveness. Programs underwent frequent changes, yet there was no tracking of whether these changes enhanced new hire productivity or retention rates. By establishing baselines and leveraging AI to analyze performance data, it became possible to discern which adjustments were effective and which constituted a waste of resources. Leaders transitioned from debating opinions to evaluating evidence.

Research supports this observation. A 2024 MIT Sloan Management Review study revealed that companies utilizing AI to analyze workforce data achieved a 30% acceleration in strategy adjustment due to the identification of correlations previously unconsidered by executives. AI can reveal actionable patterns for leaders, providing a competitive advantage.

In a specific engagement in late 2024, the implementation of AI-enhanced time tracking and forecasting resulted in significant improvements in accuracy. Executives gained real-time insight into workload and project viability, replacing debates over opinion with evidence-based discussions. Team leaders were able to more effectively assess team member availability, leading to reduced burnout and increased quality output.

Practical Application

1. Utilize AI to analyze transcripts of executive meetings and internal strategy documents.
2. Overlay training and talent data.
3. Inquire: "What patterns or correlations link our initiatives to business outcomes?"

Reflection

What is a suspected pattern within your organization (e.g., training linked to sales, engagement linked to turnover) that you have lacked the data to substantiate?

Personalizing Development

A frequently cited, yet underutilized, application of technology lies in personalizing employee learning journeys. A large global manufacturing company had historically implemented leadership programs uniformly across all regions. Over time, the limitations of this one-size-fits-all approach became apparent. By integrating AI-enabled tools, it became possible to recommend differentiated stretch assignments, micro-courses, or coaching focus areas based on role, tenure, or market. Leaders received development that was perceived as relevant, resulting in a significant increase in engagement.

The broader trend is clear. RedThread Research (2025) found that organizations employing AI for personalized learning pathways were

3.1 times more likely to report higher employee engagement and 2.7 times more likely to retain top talent. Personalization is no longer optional; it is a critical performance driver.

Furthermore, personalization extends beyond digital learning. Tracking and supporting structured innovation time revealed individuals with a natural inclination towards curiosity and experimentation. This data provided leaders with a more comprehensive understanding of individual strengths and potential risks, enabling them to tailor coaching and stretch assignments with greater precision. Personalization evolved from focusing on *what people learn* to *how they lead*.

Practical Application

1. Use AI to map employee profiles against critical skills.
2. Assign targeted learning or stretch assignments.
3. Track outcomes to assess whether personalization drives accelerated performance.

Reflection

If every leader on your team had a personalized growth plan aligned with strategic objectives, what impact would it have?

Anticipating Risks

Technology offers powerful capabilities for evaluating and mitigating risks across all organizational functions. One financial services client, for example, initially treated compliance training as a mere formality. Despite achieving 100% participation, violations persisted. By leveraging AI to analyze relevant data, they identified incident spikes in specific geographic areas where leaders had failed to reinforce the training. Armed with this insight, they adjusted their reinforcement strategies, resulting in a 30% reduction in violations and a significant decrease in regulatory exposure, saving millions. The core issue was not a deficiency in employee knowledge, skills, or abilities, but rather a systemic problem that necessitated process changes and altered management behaviors. In this instance, neither increased nor improved training would have effectively addressed the problem.

The risk management perspective is where AI frequently demonstrates its greatest value. PwC's 2025 CEO Survey revealed that 49% of CEOs anticipate a "major disruption" within the next five years, while only 24% believe their people strategies are adequately prepared. Leaders who employ AI for risk analytics can bridge this gap by proactively identifying emerging patterns.

During an efficiency review for a professional services firm that utilized billable hours, we uncovered another form of risk: rework and scope creep. By standardizing kickoff processes and implementing AI-driven templates, teams minimized costly, non-billable corrections. In this instance, the risk was not regulatory in nature; instead, it was found in the inefficiencies and errors that eroded profit margins. From an executive standpoint, this becomes a critical factor that can encourage further investment.

Practical Application

1. Utilize AI to analyze data related to compliance, turnover, and customer complaints.
2. Identify spikes or patterns that may indicate underlying risks.
3. Translate insights into actionable steps: refine training programs, adjust communication strategies, or provide targeted local support.

Reflection

What risks are of greatest concern to your executives, and how could AI facilitate earlier anticipation of these risks?

Scaling Allyship and Culture

Another underutilized application of technology involves analyzing human networks to identify change agents and informal influencers. In large organizations, the most influential individuals are not always those in senior leadership positions. AI can map these networks by analyzing communication patterns, project mentions, and feedback loops. In one instance, this approach revealed a group of mid-level managers who served as key connectors across different regions. By engaging these individuals as advocates, the client accelerated the adoption of a major transformation initiative far more effectively than through executive messaging alone.

A Harvard Business Review analysis of "networked organizations" revealed that initiatives with cross-functional advocates were twice as likely to succeed. Furthermore, Korn Ferry (2025) reported that companies leveraging AI-driven network analysis experienced 22% faster adoption of culture initiatives. My own work with various organizations corroborated this. Formalized "mindset shift" activities, such as guided quarterly conversations and innovation town halls, foster a shared language and norms. AI-supported repositories facilitate the rapid dissemination of these ideas across teams, reinforcing culture beyond formal channels. Consider the experience of joining a new company and not understanding the terminology. Once you have mastered the organization's lexicon, you begin to feel less like an outsider and your comfort level fosters stronger and broader connections across the organization.

Practical Application

1. Employ AI to scan strategy decks, meeting notes, or collaboration tools.
2. Inquire: "Which leaders or teams are most frequently referenced as critical to success?"

3. Empower these hidden influencers to become advocates for your initiatives.

Reflection

Who are the key connectors within your organization, and how are you equipping them to amplify your influence?

EFFICIENCY + EFFECTIVENESS

Leaders who prioritize automation achieve time savings, while those who embrace acceleration multiply their impact. The five dimensions of freeing capacity, surfacing insights, personalizing development, anticipating risks, and scaling culture transform AI from a back-office tool into a boardroom advantage.

Executives reward leaders who can demonstrate how their function directly drives growth, mitigates risk, and builds long-term value. AI will not perform that thinking for you, but it will sharpen your thinking and accelerate your ability to act.

∼

LEVERAGING AI AS A STRATEGIC THOUGHT PARTNER

Here's how top executives can leverage AI as a true thought partner:

1. **Utilize AI to Filter Noise from Signal.** C-suite leaders often receive an overwhelming volume of data, including investor reports, customer insights, financial models, and talent dashboards. Instead of skimming these materials, input the raw data into AI and ask: "Summarize the top three risks and opportunities identified across these sources." "What patterns emerge across customer, financial, and workforce data?"

This process transforms overwhelming data into strategic clarity.

2. **Stress-test Board Narratives.** Prior to an earnings call or board meeting, paste your draft remarks or slide deck into AI and ask: "Where are the gaps compared to industry benchmarks?" "What objections might board members or investors raise?" "How would a competitor interpret these numbers?"

AI will not replace your judgment but can anticipate challenging questions, preventing you from being caught off guard.

3. **Identify Financial and Operational Correlations.** CFOs, in particular, can use AI to analyze data beyond standard line items. Some examples are:

- "Correlate changes in attrition with revenue per employee."
- "Identify links between training investments and margin improvement."
- "Show me early signals that turnover in critical roles could impact delivery timelines."
- "Based on purchasing reports and production rates, identify the top three areas where purchasing processes can be streamlined to reduce workload or costs."
- "Evaluate these cost center reports to identify potential areas of duplicate expenses that could be consolidated."

These correlations do not prove causation; however, they highlight areas where closer scrutiny may prevent costly oversights.

4. **Expand the Leadership Lens:** C-suite leaders can leverage AI to align people strategy with business strategy:

- "Based on our last three strategy presentations, where are talent gaps most likely to impede progress?"
- "What external risks (e.g., regulatory, demographic, or technological) should we be preparing the workforce for now?"
- "What impending industry changes have I not considered?"
- "What insights can you provide regarding [insert competitive companies] that could affect us and that I may not be considering?"
- "How can we anticipate and prepare for [insert potential risk here]?"

This enables C-suite leaders to identify potential blind spots.

Reflection

If you had analyzed your last presentation using AI, what risks, correlations, or blind spots might it have revealed, and how would that have influenced your narrative?

PORTFOLIO BETS: STOP, SCALE, OR SUNSET

Technology can enhance performance, but only when leaders apply rigor to investment decisions. Often, enthusiasm or internal politics sustain underperforming initiatives beyond their useful life, while promising pilot programs are not given adequate support.

The solution is to treat each initiative as a portfolio investment, where continued funding is justified by clear leading and lagging indicators.

Leading Indicators: Fit and Early Pull

These are the early signs of success. Without them, scaling is unlikely.

- **Strategic Fit:** Does the initiative align with key strategic priorities (growth, cost reduction, risk mitigation, customer loyalty)?
- **User Pull:** Are employees, customers, or partners proactively seeking more involvement, without prompting?
- **Early Adoption:** Are pilot participants using the initiative voluntarily and providing actionable feedback?
- **Single Variable Change:** Are you testing one factor at a time to isolate the cause of observed effects? (This principle is derived from our onboarding vignette. Without baselines and focus, it is impossible to determine what drives results.)

Lagging Indicators: Sustained Outcomes

These are the quantifiable results that justify scaling or necessitate termination.

- **Retention:** Did the initiative improve customer or employee retention rates?
- **Margin:** Did the initiative protect or increase contribution margin, rather than simply increasing volume?
- **Revenue Quality:** Did the initiative generate recurring revenue (renewals, repeat customers, upsell potential), as opposed to one-time gains?
- **Resilience:** Did the initiative make processes more adaptable to disruptions (supply chain, market, or regulatory)?

The Decision Rule: Stop, Scale, or Sunset

- **Stop:** If leading indicators are absent, discontinue the initiative to avoid wasting resources.

- **Scale:** If both leading and lagging indicators are positive, increase investment.
- **Sunset:** If lagging indicators do not materialize after adequate testing, terminate the initiative. Avoid the sunk-cost fallacy.

Principle in Practice

At one organization, I applied this approach using our One Thing for One Team pilot discipline. Each pilot focused on a single variable and a single team. Success was determined by fit, pull, and outcomes, not enthusiasm or volume. Only after these criteria were met did we scale across functions or markets.

This portfolio mindset prevents diluted efforts, keeps leaders from pursuing trendy tools, and ensures that scalable initiatives generate enterprise value.

9

GOVERNING YOUR AI TOOLS

Imagine a prestigious national AI research institution, tasked with advancing technology for the public good. However, internal tensions arose, with staff anonymously reporting governance failures. Critics cautioned that the focus was shifting from societal impact to internal politics, and government funders demanded change, leading to the CEO's resignation.

This mirrors the situation at the UK's Alan Turing Institute, where in 2025, the CEO stepped down amid criticism that strategic priorities had shifted haphazardly and governance structures had failed to align mission with execution. The turmoil drew scrutiny from researchers and government officials due to the erosion of trust at the highest levels.

If an AI research powerhouse can falter not due to technical failure but due to governance shortcomings, what does that imply for organizations at all levels?

Innovation thrives with minimal oversight, but sustainable impact requires structure. AI offers speed, but without governance:

- Ideas can stray from their intended purpose.
- Biases embedded in data can go unnoticed.

- Ethical lapses can occur silently.

In the Turing case, a breakdown in alignment and oversight triggered leadership collapse. The failure occurred not because of technology failure, but because of governance deficiencies.

GOVERNANCE CREATES SUSTAINABLE RESULTS

Multiplication without structure results in chaos. AI can generate insights, correlations, and predictions continuously, but it cannot determine which ones are valuable, how they interconnect across functions, or how to ensure consistent organizational action. Many C-suite leaders mistakenly view governance as an IT issue or a compliance checklist, focusing solely on policies and data protection. While important, these aspects are insufficient. Governance is not about control, but about confidence.

For a CEO, governance means having a clear process for validating, acting on, and reporting on AI-driven correlations, such as the relationship between attrition and revenue. For a CFO, it means ensuring that AI-enhanced financial models adhere to consistent rules, preventing one division from manipulating figures while another operates with integrity. Executives must own governance, as it translates AI's potential into performance and cannot be delegated.

In one client company, different departments independently adopted AI tools. HR experimented with AI-driven recruiting screens, Finance explored AI-enabled forecasting, and Operations tested predictive maintenance. Each pilot showed promise but lacked a shared framework.

The result? Confusion reigned. Leaders received disparate figures depending on the department consulted. Some managers placed faith in the tools; others dismissed them as overhyped. When the CEO requested a consolidated report, the data failed to align. The promise of AI dissolved into unproductive noise.

Contrast this with an organization where governance was integral from the outset. They established a cross-functional council, led by the COO and supported by IT, HR, Finance, and L&D. Every AI pilot was required to directly support a business priority, adhere to a shared data model, and publish results on a central dashboard. Leaders trusted the outputs because the process was transparent and repeatable. AI empowered leaders to deliver actionable insights, driving aligned action throughout the organization.

When you, as a CEO, CFO, or CLO, establish governance for AI, you are mitigating risk, institutionalizing organizational knowledge, and creating a framework that empowers the next generation of leaders to accelerate progress, make informed decisions, and align more closely with the company's values.

Governance ensures that AI enhances both individual and organizational performance, creating a lasting legacy. Without governance, AI remains a collection of disparate tools. With governance, AI evolves into a system for sustainable competitive advantage. The leaders who master this shift will amplify their influence across the enterprise, ensuring that every insight, correlation, and innovation reinforces the company's strategic objectives and builds long-term value.

Governance is the Executive's Strategic Asset

For C-suite leaders, governance should not be perceived as a bureaucratic impediment but rather positioned as a strategic asset.

Governance ensures alignment with mission and values. Without it, even the most effective leaders can be undermined by misaligned AI tools or miscommunicated strategy.

Governance fosters trust among stakeholders. A well-governed AI system signals that insights are credible and decisions are defensible.

Governance mitigates reputational and legal risk. As demonstrated by recent corporate AI missteps, such as biased recruiting algorithms, a lack of oversight can lead to embarrassment, litigation, or more severe consequences.

Building Effective Governance

Many organizations struggle with AI governance not due to a lack of good intentions, but because their approach is reactive and siloed. Gartner projects that by 2027, 60% of AI initiatives will fail to deliver value due to fragmented governance.

In contrast, modern governance entails:

- Embedding oversight at the source ("shift-left" stewardship)
- Creating transparent data lineage
- Employing federated structures that empower teams while ensuring consistency

Governance is not mere paperwork; it is the system that provides executives with the confidence that every system and process is working in service of the business, not against it. As highlighted in an ATD article, governance serves as the "guardrails for success," reinforcing its role in enabling rather than restricting innovation. By applying governance to your technology, you can accelerate decision-making and reduce risk.

For a CEO, governance provides clarity. When AI identifies a correlation between attrition and revenue, a defined process exists to test, validate, and determine the appropriate course of action.

For a CFO, governance embodies integrity. AI-enabled models must adhere to consistent rules across divisions, ensuring forecasts are not inflated in one region while conservatively projected in another. Transparent rules safeguard credibility with investors and the board.

Without governance, leaders lack confidence in data and insights. Functional silos lead to conflicting figures, and perceived innovation devolves into unproductive noise. With governance, every decision aligns with strategy through a traceable path, yielding credible and repeatable results.

Governance is crucial for maintaining speed.

While often perceived as a hindrance, it accelerates progress by establishing standards that prevent functions from repeatedly reinventing processes. A secure foundation enables faster movement.

Most importantly, governance protects the organization's reputation. Investors and employees expect leaders to manage risks effectively, not eliminate them entirely. A robust governance system demonstrates this discipline.

Governance Provides Guardrails

I once witnessed a CHRO eagerly present a meticulously crafted talent acquisition plan to an executive meeting. Despite weeks of dedicated effort and a solid concept, the plan was swiftly dismissed due to previously unmentioned system limitations and conflicting business priorities. The plan was canceled prematurely.

This failure stemmed not from a lack of effort but from inadequate governance. A structured mechanism to integrate stakeholders from forward-looking areas like Business Excellence and Market Strategy with functions such as IT and Operations into the talent strategy was absent. Critical input surfaced too late, and the seemingly well-designed program was undermined by blind spots that effective governance should have addressed. With governance, leaders can proactively identify risks, make necessary adjustments before launch, and preserve credibility within the executive suite.

Conversely, I observed the positive impact of governance at Hilti within the Learning & Development function. Early involvement in major initiatives like the Evolution rollout or the Nuron battery launch, facilitated by robust governance, proved pivotal. A clear process connected sales, marketing, operations, and learning leaders, allowing issues to be addressed during the design phase rather than after rollout. This collaborative approach fostered a shared plan focused on mutually beneficial business outcomes, resulting in faster adoption, fewer errors, and enhanced credibility across the business.

The contrast is evident:

- Without governance, ideas fail due to a lack of stakeholder alignment.
- With governance, initiatives scale effectively because relevant perspectives are integrated from the outset.

Governance Shapes Outcomes

Sounds like an oxymoron, doesn't it? Governance and strategic paired together feels like equating tactics and strategy. However, my experience underscores that governance distinguishes between tactical training and strategic learning that drives business outcomes. At Hilti, effective governance provided learning leaders with early visibility into organizational priorities. In its absence, promising ideas stalled due to belated stakeholder objections. External research corroborates this pattern, highlighting the critical role of governance in shaping successful outcomes.

Without a robust governance structure, even well-designed programs often fail to influence strategy. Conversely, a strong governance framework enables functions to align actions with top business priorities and enhances transparency between investment and results. Rita Mehegan Smith's work in "Strategic Learning Alignment" identifies five oversight areas crucial to every governance system: accountability, operational effectiveness, program service and quality, effective controls, and adherence to enterprise priorities. These areas mirror those applied by boards of directors to corporate governance, now adapted for learning initiatives.

A 2023 study of industrial companies in Oman revealed that firms with stronger corporate governance traits—independent, well-structured boards, high audit committee quality, and clear transparency and accountability—achieved significantly better financial performance (as measured by Return on Assets, Equity, and Sales) compared to peers with weaker governance practices. In contrast, companies lacking these formal governance mechanisms often struggle with inconsistent oversight, diminished stakeholder trust, and increased volatility.

I witnessed the impact of inadequate governance at a mid-sized company where the learning team developed an ambitious leadership development program. While the design was thoughtful and the intent was positive, a lack of governance proved detrimental. The program had not been evaluated against the company's operational priorities or budgetary constraints. Consequently, during the final pre-launch approval, the CFO swiftly rejected it, citing a focus on margin recovery and a lack of evidence demonstrating the program's potential contribution. His perspective subsequently influenced all stakeholders. With effective governance, the training manager could have reframed the leadership program to address cost control, talent retention, or succession risks. Instead, the program was shelved.

Gartner's Future of HR outlook reinforces this point: fewer than 20 percent of organizations consistently leverage governance to connect talent programs to enterprise strategy; however, those that do are twice as likely to achieve a positive ROI on their learning investments.

The lesson remains consistent: governance does not impede progress; it strengthens credibility. It fosters confidence that initiatives have been thoroughly vetted against business priorities, system realities, and stakeholder needs before formal presentation.

Therefore, governance transcends mere compliance; it transforms isolated ideas into enterprise-wide initiatives, protecting leaders from wasted effort and the erosion of credibility associated with programs that fail due to belated alignment.

PRACTICAL FRAMEWORK: BUILDING EFFECTIVE GOVERNANCE

Governance does not need to be complex, but it must be consistent.

At **ansrsource**, I have observed that governance thrives when two elements are present:

1. Structured cross-functional conversations
2. A disciplined decision-making process

Without both, strategy can become misaligned. With them, team members understand that everyone operates within the same framework, accelerating action and facilitating the rapid development of talent into new roles.

Create a Rhythm of Cross-Functional Alignment

I maintain weekly meetings with our Sales VP to address the rapidly evolving revenue landscape. I also meet weekly with the Product Team to ensure continuous review of our diverse product portfolio, including SaaS, L&D as a Service, and other offerings. On a monthly basis, we convene Sales and Marketing to foster alignment between offers and outreach, recognizing that such coordination is crucial and cannot be left to chance. Furthermore, I meet weekly with the Head of Operations, monthly with the CFO, and weekly with the CEO and Founder. While each meeting provides a unique perspective, collectively, they guarantee that no initiative progresses without comprehensive cross-functional awareness.

At Hilti, this principle was instrumental in global rollouts. For instance, when launching the Evolution sales training program, we established a robust governance framework. Sales, Operations, IT, and Learning leaders engaged in regular alignment meetings, incorporating input from local advocates. This structured governance approach facilitated the early identification of potential risks and promoted smoother adoption across diverse geographical regions.

Practical Application

1. Identify your top five business drivers.
2. Schedule regular meetings with the leader responsible for each driver.

3. Maintain a consistent meeting cadence; sporadic governance undermines its effectiveness.

Reflection

Which executive conversations are you holding regularly, and which gaps in alignment are potentially impacting your credibility?

Anchor Decisions in a Clear Process

Enthusiasm alone is insufficient justification for investment. At **ansr**source, we employ a transparent decision tree, outlining the investment required, the anticipated payoff, and the scalability of the initiative. Subsequently, we develop a product requirements document that details not only the technological aspects but also the resources necessary to sustain the product in the market. Finally, the leadership team collectively evaluates the proposal. This approach transforms governance from a subjective exercise into a disciplined filtering process, enabling team members to understand the rationale behind decisions and fostering their development as more informed leaders within their respective domains.

At Hilti, we implemented a similar disciplined approach during the rollout of the Nuron battery platform. Rather than delaying training until the last minute, our governance framework mandated the evaluation of rollout sequencing, local adaptation, and sales readiness. This process proved highly effective. Time-to-market was reduced, adoption accelerated, and sales achieved their revenue targets because governance ensured the alignment of training with the launch, thereby preventing potentially costly errors.

Value Filter

Strong leaders do not reject ideas because they lack enthusiasm. They slow ideas down because they respect the cost of distraction. Distraction can come in many forms.

A value filter is the mechanism that converts enthusiasm into disciplined decision-making. It provides a visible, repeatable way to evaluate initiatives before momentum, politics, or urgency distort judgment. When applied consistently, it removes ambiguity, reduces personal bias, and reinforces enterprise thinking across the organization.

At its core, the value filter asks a small number of non-negotiable questions—always in the same order:

1. **What problem are we solving, and for whom?** If the problem cannot be articulated in business terms such as growth, cost, risk, or value creation, the initiative is not ready for investment.
2. **What is the full cost not just to launch, but to sustain?** This includes technology, people, change management, governance, and ongoing operational ownership.
3. **What outcome will justify that investment?** Activity is insufficient. The initiative must be tied to measurable impact, not projected effort.
4. **Can this scale without increasing fragility?** Initiatives that succeed only through heroics, custom workarounds, or informal knowledge do not scale, they accumulate risk.
5. **What must be true for this to succeed and what breaks if we are wrong?** Surfacing assumptions early prevents downstream failure disguised as execution challenges.

Presented visually, the value filter functions as a decision tree rather than a debate. Ideas either progress, pause, or stop based on whether they clear the same standards as every other investment.

When Politics Enter the Room

Politics often surface precisely when these questions are asked. Enthusiasm intensifies. Timelines accelerate. Leaders argue that "this time is different," or that rigor will "slow innovation." What is actually being challenged is not the process but the discomfort of transparency.

Enterprise leaders recognize this moment for what it is: *a test of discipline.*

Holding the line does not require confrontation. It requires consistency. When the value filter is applied evenly, especially to initiatives sponsored by senior leaders, it stops being perceived as resistance and starts being understood as governance.

- The fastest way to **lose credibility** is to bypass rigor under pressure.
- The fastest way to **build trust** is to apply the same standard when it is inconvenient.

Over time, teams adapt, proposals improve, and leaders begin answering the questions before they are asked. Governance shifts from gatekeeping to capability building.

Holding the Line as a Leader

Enterprise leadership is revealed in moments when enthusiasm competes with judgment.

Leaders who cave to urgency teach their teams that process is optional. Leaders who hold the filter teach their teams how to think. The goal is not to say "no" more often, it is to ensure that when the organization says "yes," it does so with clarity, commitment, and shared ownership of the outcome.

- This is how governance becomes developmental.
- This is how organizations learn.
- This is how value compounds.

Practical Application

1. Use a consistent decision tree for every major initiative, explicitly outlining investment, expected payoff, and scalability.
2. Require a requirements document that defines success beyond launch, including ownership, sustainability, and measurable outcomes.
3. Review proposals collectively with the leadership team before committing resources while focusing on trade-offs, not advocacy.

Reflection

How many of your current projects were funded based on enthusiasm alone, and what would change if every idea were subjected to a decision tree analysis?

Build a Culture of Cross-Functional Awareness

Governance does not necessarily require a formal committee. At a minimum, it necessitates that leaders understand how their decisions impact other functions and ensure that all team members are aware of the leadership's approach. At Hilti, we took this a step further by transforming awareness into active collaboration.

Several times each year, we convened business and L&D leaders to collaboratively prototype new training programs. This approach was applied to every major sales training program and to Leading Leaders, our flagship leadership program. These sessions were not protracted, bureaucratic meetings; rather, they were hands-on design workshops where business leaders tested concepts, challenged underlying assumptions, and helped shape the strategic direction.

The result was significant. Instead of L&D designing in isolation and subsequently seeking buy-in, business leaders felt a sense of ownership from the outset. L&D leaders integrated their priorities into the design process and identified potential risks early on. Consequently, program rollouts were supported by champions across the organization who were invested in the initiative because they had contributed to its creation.

Many enterprises refer to this as **matrix collaboration**, where leaders are accountable to both functional and regional or product lines, making governance even more critical. A clear cadence and a shared decision-making process transform the matrix into a strategic advantage, ensuring diverse perspectives are included and alignment is achieved rapidly across the organization.

This represents governance at its best—active involvement. When leaders see their contributions reflected in a solution, they are naturally inclined to advocate for it because it addresses their needs and priorities.

Practical Application

1. Host regular prototyping sessions that bring together cross-functional leaders.
2. Structure these as working sessions, not presentations. Emphasize that the objective is collaborative design, not mere sign-off.
3. Capture not only the desired outcomes for initiatives but also the risks and success metrics that business leaders will use for evaluation.

Reflection

When was the last time your cross-functional leaders co-created a solution with you instead of reviewing it after its completion?

Establish a Governance Committee When Appropriate

Culture and process should be prioritized initially. However, when the organization is ready, governance also requires structure. At Hilti, this structure took the form of project-specific steering committees rather than a single standing committee. Each major initiative had its own committee, deliberately composed of cross-functional representatives, including at least one challenger whose role was to question assumptions and prevent groupthink. These diverse committees also ensured that talent development was a shared responsibility and provided valuable first-hand experience in assessing emerging talent.

For L&D projects, the governance process was clearly defined, utilizing stage gates:

- Gate 0: Determine the project's viability.
- Gate 1–3: Discovery and business case development, culminating in a go/no-go decision.
- Gate 4: Prototype testing.
- Gate 5: Launch.
- Gate 6: Measure results.

This system ensured that projects remained aligned with business priorities while maintaining flexibility. It also reflected Hilti's culture of balancing support with accountability, empowering employees while upholding high standards.

While not every project benefited from this model (an interviewing training, for example, became diluted due to excessive efforts to achieve universal consensus and another, led by a passionate DEI advocate, narrowly focused on gender equity to the exclusion of other critical workforce realities), governance proved instrumental in other instances.

The launch of a social learning LMS exemplified the power of effective governance. Recognizing the influence of Hilti's permission-based culture on adoption, we partnered with Sea Salt Learning, led by Julian Stodd, to gain crucial insights. Following prototyping with the marketing team, we identified and cultivated early adopters, including Walid Hussein, who became a platform champion, and Andreas Markgraf, who expertly built and nurtured communities across the organization. Stu Ryan and Egor Semenikhin also emerged as strong advocates, leveraging their unique skills and influence to ensure the system's success. Stu focused on driving adoption among IT leadership, while Egor's technical expertise and credibility bolstered the platform's legitimacy, particularly within more skeptical departments. Together, these peer leaders transformed the LMS from a pilot program into a sustainable, enterprise-wide solution.

This illustrates the true value of governance: not as a bureaucratic impediment, but as a framework that aligns organizational culture, business objectives, and execution strategies. Well-designed governance fosters both accountability and ownership, assuring leaders that initiatives have undergone thorough vetting and empowering employees to actively shape solutions. Ultimately, it prevents the costly waste of launching programs that lack fundamental stakeholder buy-in.

Practical Application

1. Prioritize project steering committees as a foundation before implementing enterprise-level governance.
2. Incorporate a designated "challenger" role to proactively identify and address potential blind spots.
3. Utilize stage-gate processes to maintain disciplined decision-making throughout the project lifecycle, from initial concept to final measurement.

Reflection

If your next initiative were subjected to a stage-gate system, at which point would it likely stall, and what insights would this reveal regarding its overall readiness?

GOVERNANCE PREDICTS SUCCESS

Research substantiates governance as a key predictor of successful execution, with cross-functional involvement serving as a significant multiplier. A 2024 McKinsey study revealed that transformation projects employing cross-functional governance were 2.5 times more likely to meet or exceed their objectives. At Hilti, this finding translated into tangible results. Champions like Walid Hussein infused marketing energy, Andreas Markgraf cultivated thriving communities, Stu Ryan spearheaded adoption among technology leaders, and Egor Semenikhin applied his technical acumen to enhance the platform. The governance structure transformed a social learning LMS from a pilot into a sustainable enterprise platform. While research elucidated the underlying principles, the dedication and expertise of our people made it a reality.

However, diverse representation alone is insufficient. The inclusion of challenger roles is crucial to mitigate blind spots. Deloitte's 2025 Global Human Capital Trends report underscores this point, noting that organizations incorporating formal challenger roles in their governance committees experienced 30% fewer failed initiatives. Hilti's consistent practice of appointing challengers mirrored this finding, resulting in projects that addressed critical questions early and yielded more robust outcomes.

Beyond mitigating failure, governance enhances adoption rates and return on investment. A 2024 Harvard Business Review study reported that organizations with structured governance processes experience 20–40% faster adoption of new systems and a 25% higher ROI on people-related initiatives. This was demonstrated at Hilti, where the adoption of the social LMS and Salesforce Evolution spread rapidly and remained effective due to structured stage gates and empowered champions. Conversely, weaker governance, such as in the diluted interviewing training or the narrowly focused DEI initiative, resulted in diminished impact.

The overarching pattern is clear: governance is not bureaucratic red tape; it is a strategic multiplier. By incorporating diverse cross-functional perspectives, designated critical reviewers, and structured processes, organizations enhance credibility, accelerate adoption, and achieve superior outcomes. This is supported by research and validated by Hilti's experience.

~

LEVERAGING AI AS A STRATEGIC PARTNER

Governance hinges on clarity, alignment, and accountability. AI can expedite these elements, provided it is utilized as a strategic partner, not as a substitute for human judgment. You remain the strategist, while AI sharpens your inquiries, challenges your assumptions, and identifies blind spots that your teams may be hesitant to address. AI needs to be your thought partner, not your thought leader. Additionally, you always have to doublecheck the output from any AI tool.

Here's how to leverage AI to enhance your governance role:

1. **Assess Alignment:** Upload your most recent CEO or CFO address, quarterly board presentation, or investor briefing and ask AI:

   - "Which initiatives and metrics did I emphasize most prominently?"
   - "How clearly do these align with our stated strategic objectives?"
   - "Where might my messaging appear disconnected from the day-to-day experiences of employees?"

2. **Map Stakeholders and Influence:** Input the last three sets of leadership meeting minutes into AI and ask:

- "Which leaders or teams do I consistently identify as critical to our success?"
- "Who appears underrepresented in these discussions, and how might this silence pose a risk?"

3. **Simulate Governance Meetings:** Provide AI with a concise description of a recently approved initiative and ask:

- "Assuming the role of the COO, what key questions would you pose regarding potential execution risks that I may have overlooked?"
- "Assuming the role of the CLO, what aspects of my framing of the initiative's people-related components might be unclear?"
- "Where am I using language that could potentially impede the achievement of other strategic priorities?"

4. **Identify Emerging Risks:** Overlay your strategic plans with external reports, including industry trends, regulatory changes, and competitor activities, and ask AI:

- "What emerging risks are not adequately addressed by our current priorities?"
- "How might my messaging unintentionally downplay or contradict these risks?"

Reflection

If AI were to analyze the transcripts of your last three leadership updates, what contradictions or ambiguous signals would it identify, and how might these be undermining the trust and clarity essential to your governance system?

～

THE EXECUTIVE OPERATING SYSTEM

Effective governance requires practical application. Leaders need more than just values or guidelines; they require a consistent operational rhythm. This rhythm is a method of managing the business that balances structured discipline with necessary adaptability.

In my experience as Chief Innovation Officer, I've found that a consistent operating rhythm is essential. Without it, decision-making becomes delayed, initiatives lose focus, and strategic direction becomes unclear. Conversely, a well-defined rhythm sharpens alignment, ensures resources are allocated to the right priorities, and accelerates execution.

I recommend the following Executive Operating System, which provides a structured cadence of conversations and decisions to keep leaders focused on key objectives.

Weekly

- **Sales + Product Rhythm:** A weekly meeting to align market feedback with product development. Key discussion points include customer requests, high-traction features or offers, and adjustments needed for current initiatives (acceleration, deceleration, or re-scoping).
- **Operations Sync:** A review of key delivery, service, and supply chain metrics to ensure efficient, high-quality performance. This meeting identifies bottlenecks where minor adjustments can improve margins or increase throughput.

Monthly

- **CFO Alignment:** A deep dive into financial performance, focusing on the implications of the numbers. The discussion goes beyond simply reporting results to identifying data-driven decisions, such as which initiatives to fund and which to sunset to avoid resource depletion.

- **Marketing + Sales Integration:** A joint review of pipeline health, examining conversion rates, opportunity quality, and the impact of marketing investments on sales velocity.

Quarterly

- **Cross-Functional Review:** A meeting of Sales, Product, Operations, Finance, and People leaders to assess progress against the top 3–4 enterprise priorities. The discussion focuses on evidence of success and necessary pivots.
- **Gate Reviews:** A stage-gate process (adapted from Hilti) where projects move through defined stages: Gate 0 (idea approval), Gate 3 (go/no-go decision after discovery), Gate 5 (launch readiness), and Gate 6 (results measurement). This process ensures that projects are evaluated based on evidence and that promising initiatives receive adequate support.

Benefits of This System

This simple yet powerful cadence:

- **Creates Predictability:** Leaders know when decisions will be made.
- **Increases Transparency:** Evidence is visible across functions.
- **Reinforces Discipline:** Projects must demonstrate value to progress.
- **Builds Adaptability:** Frequent touchpoints surface risks and opportunities early.

By adopting a structured operating rhythm, leaders can shift from reactive problem-solving to proactive strategic direction, setting an example of effective leadership for their teams and building a lasting legacy.

WORKSHEET: OPERATING RHYTHM

Purpose: To design an executive operating system that ensures decisions are made at the right level, with the right evidence, at the right time.

Cadence	Who is in the room?	What decisions?	What evidence required?
Weekly	e.g., Sales VP + Product Head	Product backlog priorities, sales enablement focus	Customer feedback, pipeline trends, adoption data
Weekly	e.g., COO + Ops leaders	Process bottlenecks, resource allocation	Throughput metrics, quality data, customer complaints
Monthly	CFO + Function Heads	Budget reallocations, stop/scale/sunset calls	Variance analysis, initiative ROI, forecast accuracy
Monthly	Sales + Marketing	Campaign alignment, pipeline quality	MQL→SQL conversion, cost per opportunity, win rates
Quarterly	Executive Team	Enterprise priorities, pivots, governance reviews	Top KPIs, risk scans, talent pipeline status
Gate Reviews	Project Teams + Sponsors	Move/Hold/Kill Project	Gate 0–6 Deliverables, Business Case, Adoption Data

10

LEAVE A LEGACY OF LEADERSHIP

L eadership legacy is not an abstract concept. It is embodied in the next generation of leaders who can make difficult decisions under pressure, maintain focus on the broader vision amidst competing details, and confidently advance transformation. It encompasses the systems and mindsets you establish that empower others to achieve greater progress, more rapidly. Every action you take serves as a model for your team members.

This chapter focuses on developing the leadership pipeline as your legacy. The next chapter will address what enterprise leadership must become systemically; this chapter focuses on the personal responsibility leaders have to prepare others to lead before that future arrives. It is about scaling your influence through people by equipping them to think strategically, act decisively, and transform organizations long after your departure.

I once worked with a regional vice president who was measured, respected, and quietly influential. While he may not have been the most flamboyant leader, lacking fiery speeches or dazzling breakthrough ideas, his impact became undeniably clear over time.

Leaders who had previously worked under his guidance were now managing major functions across the company. One was spearheading sales in a new market, another was driving operations in a region that had doubled revenue, and a third was selected to join the global leadership team. When describing their leadership style, these individuals consistently used terms such as "calm under pressure," "strategic," and "disciplined." These were not coincidental traits; they were qualities that the VP had consistently modeled and cultivated.

His true strength lay not in his individual achievements, but in the leaders he developed. He didn't simply instruct his team on what to do; he taught them how to think. He guided them through complex decisions, explaining the trade-offs involved. He challenged them to look beyond daily tasks and consider the larger strategic context. He involved them in transformation projects, ensuring they understood not only the "what" but also the "why" of organizational change.

Upon his retirement, his legacy was not defined by a list of completed projects but by a pipeline of leaders capable of guiding the business long after he was gone. His influence was pervasive, even when unacknowledged.

This is the essence of legacy leadership: not programs, not processes, not short-term wins, but the cultivation of individuals who can carry the mission forward because you have equipped them to do so.

WHY LEGACY MATTERS

Most leaders are evaluated based on metrics such as revenue growth, margin improvement, and efficiency gains. While these measures are important, their impact is often fleeting. A quarterly target is soon forgotten, a product launch becomes outdated, and even the most celebrated transformation loses its luster as new challenges emerge.

The enduring element is people: the leaders you develop, the judgment you help them cultivate, and the confidence you instill in them to make decisions in uncertain times. These are the true hallmarks of leadership.

Legacy matters because:

1. **It provides adaptability in a dynamic environment.** In today's rapidly evolving business landscape, industries are subject to unprecedented change. No single leader can navigate every disruption in isolation. Developing a pipeline of leaders who possess shared values and sound judgment is crucial for organizational resilience.

2. **It ensures you amplify your influence through leadership development.** Cultivating leaders who emulate your clarity of thought and action extends your influence beyond your direct role. Your principles and mindset are perpetuated throughout the organization, ensuring a consistent approach even in your absence.

3. **It preserves culture and strategy through deliberate development.** Organizational culture and strategic alignment can erode without intentional leadership development. New hires and promotions may lead to deviations from established norms. By investing in training that explains both the rationale and process behind key decisions, you reinforce cultural values and strategic discipline.

4. **It enables organizational transformation through distributed leadership.** Transformation is rarely a singular event; it requires a cascade of leaders at all levels who are equipped to guide others through ambiguity. A robust leadership pipeline ensures that transformation efforts are sustained, becoming ingrained in the company's operational fabric, rather than collapsing with the departure of a key executive.

For leaders in support functions such as L&D, HR, IT, or Finance, this is particularly vital. Your value lies not only in your technical expertise but also in your capacity to develop leaders who can transcend functional silos. You are developing leaders who balance accountability with empathy, who understand the interconnectedness of the business, and who are prepared to assume greater responsibilities when the organization requires them most.

Legacy is a personal strategic imperative for leaders, not an organizational abstraction.. As you approach retirement, resignation, or transition to new opportunities, the business will continue. The salient question is whether your leadership development efforts have positioned the organization for enhanced success.

A LEGACY OF COURAGEOUS LEADERSHIP

Throughout my career, I have had the privilege of collaborating with Terry Copley, a leader whose name carries significant weight within Hilti, my previous employer. His reputation is not solely based on his title but on the caliber of leaders he has developed.

Terry is a leader who leaves a lasting impression. He empowers others, coaches with conviction, and inspires individuals to exceed their perceived limitations. While his approach may not always be gentle as he can be impatient with those lacking confidence or business acumen, those who embrace his guidance find their potential limitless.

His legacy extends beyond specific programs or initiatives, though he has a proven track record of successful projects. His true legacy is the multitude of leaders who bear his imprint: courageous in decision-making, empathetic yet firm, and focused on performance without compromising the well-being of their teams. He exemplifies kindness over mere niceness, demonstrating that leadership involves delivering honest feedback with compassion rather than avoiding difficult conversations. He leverages storytelling to inspire and to convey complex concepts in a universally accessible manner.

Terry's influence extends far and wide, rooted in his personal approach. Individuals at Hilti readily share anecdotes of his encouragement, the challenges he posed, and the unwavering support he offered during critical moments. His charisma and conviction would have made him an exceptional advocate for any cause, inspiring unwavering belief in those around him.

What distinguishes Terry is his evolution. While the seeds of his leadership style were present early on, it was in his 40s that he fully embraced his potential. Since then, he has dedicated himself to leaving a lasting impact, fostering a generation of leaders characterized by boldness, compassion, and effectiveness.

Terry's story serves as a powerful reminder that legacy is not a product of position but rather a reflection of how one treats, challenges, and prepares others to lead. His example underscores that a legacy of leadership can be the most significant contribution any executive makes.

PRACTICAL FRAMEWORK: BUILDING A LEADERSHIP LEGACY

Legacy is not accidental; it stems from intentional choices leaders make daily, encompassing their focus, engagement of others, and balance between results and people. Here are five practices to ensure your leadership endures through others:

Model Decision-Making, Not Just Decisions

Many leaders shield their teams from difficult choices, announcing decisions without revealing the underlying process. This fosters dependency, not leadership.

Practical Application

1. Incorporate team members into decision-making discussions.
2. Discuss trade-offs: the considerations, acceptable risks, and rationale.
3. Demonstrate your thought process, not just your conclusions.

Reflection

What recent decision could have served as a valuable learning opportunity for your team?

Teach Big-Picture Thinking

Legacy-driven leaders elevate vision, not just delegate tasks. They help individuals understand how their function aligns with strategy, customers, and the broader market.

Practical Application

1. Use recurring meetings to connect departmental work to organizational goals.
2. Ask your team: "How does this project contribute to revenue, cost management, risk mitigation, or growth?"
3. Encourage cross-functional shadowing or developmental projects.

Reflection

How often do you assist your team in connecting their work to the overarching business strategy?

Stretch Without Breaking

Effective leaders provide challenging opportunities that build confidence and capability, avoiding both easy wins and insurmountable obstacles. Terry Copley excelled at this, consistently pushing individuals beyond their perceived limits without setting them up for failure.

Practical Application

1. Assign projects that slightly exceed an individual's current skill set.
2. Pair them with a mentor or peer for guidance.
3. Recognize progress over perfection.

Reflection

Who on your team is prepared for a greater challenge, and how can you facilitate it?

Build Courage And Empathy Together

Kindness, rather than mere niceness, distinguishes legacy leaders. While niceness may avoid conflict, kindness delivers truth with consideration. Accountability paired with empathy fosters growth.

Practical Application

1. Provide feedback that is direct, specific, and respectful.
2. Model ownership of mistakes and subsequent learning.
3. Demonstrate care in how accountability is delivered, not only in its delivery.

Reflection

Does your team perceive you as both demanding and supportive, or is only one aspect evident?

Multiply Through Storytelling

Stories endure beyond instructions. Leaders who frame experiences as narratives cultivate shared memory, which in turn shapes culture.

Terry's effectiveness stemmed from his ability to transform lessons into memorable and repeatable narratives.

Practical Application

1. Transform a project, failure, or success into a teaching narrative.
2. Share stories illustrating values in action that yield desired outcomes.
3. Encourage team members to share their own learning experiences through storytelling.

Reflection

What narrative from your leadership journey are you conveying and what story are your people recounting about you?

∾

LEVERAGING AI AS A STRATEGIC THOUGHT PARTNER

AI can serve as a mirror for leaders who want to understand whether their influence truly extends beyond their presence.

1. **Upload Your Own Voice.** Input transcripts from your three most recent executive presentations, town halls, or board updates into an AI platform and pose the following questions:

- "Which themes, priorities, or phrases do I emphasize most frequently?"
- "What aspects might be challenging for a first-line manager or director to comprehend?"

- "Where might my message contain inconsistencies or contradictions with previous statements?"

2. **Overlay Team Readiness.** Provide the AI tool with concise profiles of your direct reports (excluding sensitive data), including their roles, current projects, and development objectives. Then, ask:

- "If I were unavailable for six months, which of these leaders could effectively champion the themes I emphasize?"
- "Where are the most significant gaps in confidence or capability that would hinder them from effectively representing my voice?"

3. **Stress Test Your Legacy.** Instruct the AI tool to role-play as the board or CEO (if you are a functional leader) and assess your team's readiness:

- "What three questions would the board likely pose to this leader if they were substituting for me?"
- "Where would their responses fall short in addressing growth, cost, risk, or culture?"

4. **Find the Blind Spots.** Challenge the AI tool to provide candid feedback:

- "What aspects of my communication style could impede successors from effectively conveying my message?"
- "What habits or deficiencies might undermine my intended legacy?"
- "What assumptions are you making and what might change if your assumptions are wrong?"

11

THE FUTURE OF LEADERSHIP
IS ENTERPRISE LEADERSHIP

L egacy endures when capability outlives the individual. In the previous chapter, we examined leadership legacy as a personal responsibility: developing people who can exercise judgment, act with courage, and lead with empathy long after you step aside. But personal legacy alone is not sufficient. For leadership to endure at scale, it must be embedded into how the organization operates.

Enterprise leadership is the mechanism that makes legacy repeatable.

It is not defined by title, charisma, or functional excellence. It is defined by the systems leaders build that cause sound judgment, disciplined decision-making, and enterprise thinking to become the organizational default. People are the legacy; systems are the carrier that preserves and scales their capability.

This book has been building toward this point deliberately. We began by dismantling the myth that influence comes from merely earning a "seat at the table." We reframed leadership from delivery to transformation, anchored credibility in business acumen, and expanded leadership from individual performance to organizational capacity. We examined how leaders translate activity into results, cultivate allies, shape culture, and govern with discipline. We explored how technology and AI can amplify leadership when integrated thoughtfully into people-centered systems. We then defined legacy as the daily discipline of preparing others to lead in your absence.

The final step is institutionalization. Contemporary leadership extends beyond functional mastery or isolated leadership skills. It requires enterprise leadership: the ability to transcend departmental boundaries and deliberately shape the organizational system itself. This chapter challenges you to convert everything you have built into an operating model that endures beyond you.

The objective is not better presentations or more persuasive narratives. It is a superior leadership operating system, characterized by:

- **A disciplined cadence** of quarterly business reviews, monthly talent discussions, and weekly execution huddles that consistently link functional activity to growth, cost management, risk mitigation, and value creation.
- **Explicit decision rights** that clarify cross-functional trade-offs by defining who decides, who contributes, and when escalation is required.
- **Shared metrics and a unified "results spine"** that enable leaders across functions to communicate in a common enterprise language.
- **Succession scaffolding** that prioritizes capability and judgment over personality or tenure, ensuring continuity through change.

For leaders in enabling functions such as L&D, HR, IT, Finance, or Operations, legacy is realized when the function continues to articulate enterprise value without you in the room. The true measure of leadership is not personal influence, but organizational fluency.

Your task, therefore, is to institutionalize the translation habit. This requires durable mechanisms such as templates, review gates, operating rhythms, and governance processes that make the question *"What impact will this have on growth, cost, risk, or value?"* unavoidable. When that question becomes reflexive rather than rhetorical, enterprise leadership has moved from aspiration to infrastructure.

ENTERPRISE LEADERSHIP

The virtual boardroom was full. Quarterly results had exceeded expectations, and the agenda reflected ambition: expansion into a new market, digitization of a core process, and the integration of a recent acquisition. What distinguished the meeting, however, was not the quality of the ideas presented, but the system of thinking on display.

Each function arrived prepared to report upward. Only one arrived prepared to connect laterally.

This distinction matters. Enterprise leadership does not emerge when individuals perform well within their lanes; it emerges when translation becomes a shared discipline across functions rather than a personal performance skill.

Around the table, executives spoke from their respective perspectives. The CFO outlined financial projections. The COO highlighted operational risks. The CHRO raised concerns about leadership capacity. The CTO described the technology roadmap. Each contribution was competent. Each was incomplete in isolation.

Then one leader altered the conversation, not by asserting authority, but by synthesizing the system. Rather than advocating for their function, they asked questions that connected the whole:

- "If we accelerate into this market, do we have the leadership bench strength to sustain it?"
- "If we digitize this process, are we prepared to retrain the workforce for new ways of working?"

- "If we integrate this acquisition, how do we protect our culture while improving margin?"

In that moment, the conversation shifted from updates to judgment. Growth, cost, risk, and culture were no longer competing narratives; they became a single, integrated enterprise story. This is the inflection point where translation moves from individual capability to organizational default.

If every quarterly review, project brief, and dashboard update followed this same pattern of explicitly linking growth, cost, risk, and culture, the organization would begin to think and speak in enterprise language by design. Enterprise leadership is not charisma at scale; it is translation embedded into systems.

When the meeting adjourned, the CEO approached the leader and said, "That's what I need from my leaders: enterprise leadership, not just functional updates." CEOs rarely mean "be more visionary." What they are asking for is a system that consistently produces this level of thinking, regardless of who is in the room.

Enterprise leadership is therefore visible not in speeches, but in structure. It shows up in how agendas are designed, how success is measured, and how teams are prepared for decision-making forums. When templates, dashboards, and operating rhythms all reinforce the language of results, behavior follows. Culture aligns not through slogans, but through repetition.

This is the trajectory of leadership in the years ahead. Titles—CLO, CHRO, CIO, CTO, CFO—matter far less than the ability to step beyond functional dashboards and steer the organization as a whole. Enterprise leaders do not protect lanes; they design the roadway.

The ultimate test is durability. When successors can step into the system, ask the same questions, and exercise the same quality of judgment, enterprise leadership has moved from person to process. At that point, translation becomes institutional memory and legacy becomes repeatable.

Enterprise leadership, however, is not proven in moments of stability. It is tested when assumptions break, timelines compress, and trade-offs become unavoidable. In periods of disruption, the question is no longer whether leaders understand the enterprise, but whether the system they have built can continue to think, decide, and act coherently under pressure. This is where enterprise leadership either reveals its strength or exposes its absence.

Disruption is Everyone's Business

Enterprise leadership is not proven in periods of stability. It is revealed under pressure.

Organizations today face disruptions that cut across every function simultaneously. Competitive threats do not respect organizational boundaries. Cultural fractures do not remain contained within HR. Supply chain failures, regulatory shifts, cybersecurity incidents, and workforce disruptions demand coordinated response, not isolated fixes. Disruption tests not only leaders, but the systems they have built to think, decide, and act when conditions change.

Organizations that treat enterprise leadership as a core capability recover faster because they have clearer operating logic. Shared dashboards, explicit decision rights, and established escalation paths allow leaders to integrate growth, cost, risk, and culture in real time. These structures reduce friction precisely when speed and coherence matter most.

In disruption, translation becomes the connective tissue of the enterprise.

When translation is embedded, finance understands the risk narrative behind operational decisions. HR grasps the cost and capability implications of growth strategies. Technology aligns investment priorities to business outcomes rather than feature delivery. Without this shared framing, functions respond independently, optimizing locally while the enterprise fragments globally.

This is why enterprise leadership is no longer optional. The leaders who will thrive in the coming decade are those who can connect growth, cost, risk, and culture and who deliberately cultivate others to do the same. When enterprise thinking is concentrated in a few individuals, organizations become fragile. When it is distributed through systems, they become resilient.

Leadership is reinforced through what is celebrated and what is tolerated. Behaviors that are rewarded are repeated; behaviors that are ignored are implicitly endorsed. Over time, these signals harden into norms. High-performing organizations embed enterprise signals into their systems—performance reviews, budget cycles, talent discussions, and technology investments—so alignment does not depend on memory or intention. When enterprise logic is reinforced through operating rhythm, culture remains coherent even as strategies evolve.

This book has outlined the building blocks required to make that coherence possible. Business acumen establishes credibility. Talent development multiplies leadership capacity. Speaking the language of results converts activity into value. Allies extend influence beyond the room. Governance creates discipline. Culture sustains momentum. Technology and AI accelerate insight when aligned to people and purpose. Legacy ensures that capability endures.

Enterprise leadership becomes self-sustaining when:

- Every plan is framed through growth, cost, risk, and culture.
- Every meeting reinforces enterprise questions, not functional updates.
- Every review links people, performance, and outcomes.
- Every successor is taught how to think systemically, not just execute tasks.

When these conditions exist, organizations no longer depend on translators to reconcile competing narratives. Enterprise fluency becomes the default.

Every decision, therefore, is constructive or corrosive. Each choice shapes culture, develops or erodes talent, and strengthens or weakens the enterprise. Leadership is not neutral. What you build today determines whether the organization becomes more resilient or more brittle tomorrow.

This mandate extends beyond any individual role or tenure. Leaders are called to build organizations that can deliver both profit and purpose, even as leadership changes hands. Gonzalo Martinez illustrates this principle well. When he participated in the IMPACT program, he showed curiosity and drive but was still forming his leadership voice. Today, as General Manager of Schaefer Brush, he shapes strategy and develops others using the same enterprise principles he learned earlier in his career. His success is not evidence of a program's effectiveness alone, it is proof that investing in people compounds long after direct involvement ends.

Disruption will continue. The question is whether the systems you leave behind will fragment under pressure or rise to meet it.

PRACTICAL FRAMEWORK: IMPLEMENTING ENTERPRISE LEADERSHIP

Enterprise leadership is not expressed through occasional strategic insight; it is revealed through daily decisions, recurring conversations, and the behaviors an organization consistently rewards/celebrates or tolerates. The practices that follow are not leadership techniques; they are **maintenance routines for the enterprise system.**

When these routines are embedded, translation persists even as priorities shift and leaders change. Enterprise leadership endures when habits outlive individuals. Without these routines, strategy decays into intention, and influence erodes as soon as attention shifts elsewhere.

Connect the Dots, Not the Silos

Enterprise leaders begin with a single, non-negotiable question:

"How does this decision affect the enterprise as a whole?"

By deliberately integrating growth, cost, risk, and culture into a unified discussion, leaders avoid fragmented decisions that optimize one function while weakening the broader system. Enterprise thinking replaces local optimization with systemic judgment.

Organizations that do this well do not rely on individual brilliance; they hardwire this question into their governance cadence:

- **Quarterly:** Cross-functional reviews that explicitly link every major project to the enterprise scorecard.
- **Monthly:** Joint checkpoints for value creation and risk emergence.
- **Weekly:** Brief alignment huddles that surface early indicators.

When this cadence is consistent, translation becomes an operating discipline not a situational skill.

- **Build** → Integrate decisions across functions.
- **Destroy** → Defend departmental metrics in isolation.

In one global manufacturing firm, the head of IT proposed a significant system upgrade. When framed as a technology investment, the proposal met resistance. Operations leaders reframed the discussion by connecting the upgrade to a three-day reduction in order fulfillment time that would directly improve revenue velocity and customer satisfaction. Once the decision was translated into enterprise outcomes, alignment followed. The system was no longer a technology project; it became a business driver.

This shift did not occur because one leader argued more persuasively. It occurred because the organization used a shared enterprise lens to evaluate the decision.

Practical Application

1. In your next executive meeting, frame your contribution by explicitly connecting it to at least one other function's priorities.
2. Model enterprise thinking for your team by asking them in project reviews: "How does this impact finance, customers, or operations?"

Develop Leaders Who Can Replace You

Enterprise leadership is measured by what happens when you are not in the room. The durability of any leadership system depends on whether the next generation can operate across boundaries using shared enterprise logic.

The objective is not to produce successors who replicate your style, but leaders who can fluently navigate the organization's operating system by connecting growth, cost, risk, and culture without requiring translation from above. Sustainability is achieved when enterprise judgment is distributed, not centralized.

Enterprise leaders therefore treat succession as a system, not an event. They conduct structured bench reviews on a regular cadence, assess potential successors using shared scorecards, and design development experiences that connect emerging leaders to enterprise outcomes rather than narrow departmental targets. Mentorship, in this context, is not advisory, it is architectural.

- **Cultivate:** Stretch individuals beyond their comfort zone; coach them in cross-functional thinking.
- **Inhibit:** Hoard knowledge, protect turf, or delay succession conversations.

At Hilti, a sales manager identified potential in a frontline account manager and deliberately assigned him a cross-functional project with marketing. The assignment exposed him to pricing strategy, customer segmentation, and demand generation, which were capabilities far beyond his immediate role. Within three years, he advanced to regional director. More importantly, the manager had built leadership capacity into the system rather than anchoring it to himself. Immediate results were achieved, and a future pipeline was strengthened simultaneously.

This is enterprise leadership in practice: development that serves both present performance and future resilience.

Practical Application

1. Identify two individuals on your team with the potential to operate at an enterprise level. Assign each a cross-functional initiative and coach them explicitly on how their decisions affect growth, cost, risk, and culture.

Reward the Right Behaviors

Reward systems are not neutral. They encode priorities and teach the organization how to behave when trade-offs are real. In enterprise leadership, recognition functions as a form of governance that signals what the system values when pressure is applied.

Leaders who reward only short-term results, regardless of how they are achieved, inadvertently institutionalize risk. Enterprise leaders reward courage, collaboration, and accountability because these behaviors preserve long-term value. Over time, consistent recognition becomes instructional: it teaches the organization how to translate decisions into enterprise outcomes.

High-performing organizations standardize recognition by explicitly linking behaviors to business impact. Success stories are not shared for inspiration alone; they are structured to reinforce enterprise logic by connecting actions to growth, cost, risk, and culture. In this way, performance management becomes a cultural language laboratory, shaping how leaders think, decide, and act.

- **Cultivate:** Publicly recognize leaders who demonstrate enterprise values under pressure.
- **Inhibit:** Tolerate toxic performers solely based on their numerical results.

A financial services firm provides a clear illustration. During a compliance review, a manager identified irregularities that could have been concealed to avoid scrutiny. Instead, she escalated the issue, triggering short-term disruption. Rather than penalizing the delay, the CEO publicly commended her decision, signaling that integrity and risk stewardship outweighed convenience. That single act of recognition recalibrated expectations across the organization: how results are achieved mattered as much as the results themselves.

This is enterprise leadership in action. Recognition becomes a control mechanism by shaping judgment.

Practical Application

1. In your next recognition opportunity, articulate explicitly **how** the outcome was achieved and **which enterprise values** were reinforced.
2. Ask yourself: "What behavior am I tolerating today that weakens our culture tomorrow?" Then act accordingly.

Leverage Technology To Multiply People

Technology is an amplifier. It multiplies whatever leadership intent already exists.

Enterprise leaders do not pursue AI, systems, or platforms for novelty or efficiency alone. They deploy technology deliberately to expand human judgment, free capacity for higher-value work, and reinforce enterprise thinking at scale. When misapplied, technology accelerates fragmentation. When aligned, it hardwires translation into the organization's operating system.

The most effective enterprise leaders use technology to *teach the system how to think*. Dashboards, AI assistants, and workflow tools are designed to surface enterprise relevance by automatically signaling how initiatives connect to growth, cost management, risk mitigation, and value creation. In this way, data becomes instructional, shaping decision-making long before escalation is required.

- **Cultivate:** Align technology with culture and people development.
- **Inhibit:** Force adoption without context or use technology as a substitute for leadership.

A healthcare organization illustrates this distinction. Leaders introduced AI-driven scheduling software to improve efficiency. Initial resistance was strong; staff feared automation would erode professional autonomy. Rather than mandating adoption, leaders reframed the technology's purpose around what clinicians valued most: increased time with patients. By demonstrating how AI reduced administrative burden and restored time for direct care, adoption increased rapidly. The system did not replace judgment; it protected it.

The result was not just better schedules, but stronger alignment between technology, culture, and mission. AI became an enabler of human-centered performance rather than a source of disruption.

Enterprise leadership demands this discipline. Technology should reduce cognitive load, surface insight, and reinforce enterprise priorities. When systems are designed with intent, they elevate people instead of sidelining them.

Practical Application

1. Before approving a technology initiative, consider: "How does this support our people and prepare them for future challenges?"
2. Pilot new tools in contained environments, observe adoption patterns closely, and adjust deployment based on how effectively the technology reinforces enterprise thinking.

Build A Lasting Legacy

Enterprise leadership culminates in the ability to leave behind a system that continues to make sound decisions long after individual leaders move on. Legacy, at this level, is not sentiment or reputation; it is the durability of judgment embedded in the organization.

Every decision contributes to that durability or erodes it. Each choice either strengthens enterprise resilience or introduces vulnerabilities that compound over time. Leaders who understand this treat decision-making itself as an asset to be cultivated.

Sustained enterprise leadership therefore requires a disciplined feedback loop. High-performing organizations institutionalize regular reflection where leaders evaluate decisions through a simple but powerful lens: **Did this decision build the enterprise or weaken it?** Over time, these reviews create a living archive of organizational judgment: a pattern library of decisions, trade-offs, and consequences that reveal what strengthens or destabilizes the system.

When captured consistently, this archive becomes intelligence. AI tools can be trained on these decision patterns to improve the quality of future choices. In this way, translation evolves from a leadership behavior into organizational DNA.

- **Build** → Act with consideration for both immediate results and the development of future capability.

- **Destroy** → Optimize short-term gains at the expense of long-term organizational health.

A consumer goods company illustrates this discipline well. Leaders faced an opportunity to significantly reduce costs by outsourcing production overseas. While the short-term savings were compelling, they paused to evaluate broader implications: the impact on workforce capability, community trust, and brand reputation. Instead of pursuing immediate margin relief, they invested in modernizing domestic facilities and upskilling their workforce. Financial returns took longer to materialize but employee loyalty increased and the company strengthened its reputation for responsible growth. This decision built enterprise capacity.

This is legacy at the enterprise level: decisions that prepare the organization to perform across generations.

Practical Application

1. At the end of each week, reflect on: "What did I build this week that will have a lasting impact?"
2. Ask your direct reports the same question and listen for patterns.

∼

LEVERAGING AI AS A STRATEGIC THOUGHT PARTNER

In earlier chapters, AI has been positioned as a strategic thought partner for individual leaders—supporting reflection, sharpening judgment, and improving translation. At the enterprise level, that role evolves. Here, AI functions not as a personal assistant, but as a **system-level mirror:** reinforcing enterprise priorities, surfacing misalignment, and preserving translation across leadership transitions.

In mature organizations, AI can be integrated into the governance framework to help sustain enterprise coherence. When embedded into quarterly reviews, operating rhythms, or leadership development processes, AI serves as a **continuous auditor** evaluating whether communication, decisions, and investments remain aligned with the organization's priorities for growth, cost management, risk mitigation, and value creation.

Used in this way, AI does not replace leadership judgment. It reflects it back consistently, dispassionately, and at scale.

The following prompt is not intended as a one-time exercise. It is designed to function as a **recurring pulse check**, ensuring that the leadership system remains fluent in the language of results and coherent across functions. AI becomes a mirror for the enterprise narrative leaders are reinforcing, intentionally or not.

Enterprise Translation Audit (AI Prompt)

1. **Feed the Context – Individually and Systemically**

- Compile the last three executive presentations, CEO addresses, and board updates, including your own and those from peer functions. Add business reviews, talent dashboards, and customer reports. Upload these documents together to enable AI to analyze the organization's overall narrative, rather than isolated perspectives.

2. **Ask AI to Play Back Your Leadership**

- What consistent themes, priorities, or trade-offs appear across these communications?

- If leadership changed tomorrow, what would the organization believe truly matters based on this language?
- Where do stated priorities align or conflict across functions?

3. Conduct a Deeper Analysis

- What ambiguous messages might teams be interpreting from these communications?
- What assumptions about growth, cost, risk, or culture are being reinforced implicitly?
- Where does the system reward activity over outcomes?

4. Evaluate the Future Impact

- If this narrative persists over the next three years, what are the likely effects on enterprise performance, leadership capability, risk exposure, and culture?
- Where might second-order consequences emerge?

5. Initiate a Strategic Shift

- What single change to messaging, decision framing, or operating rhythm would most improve enterprise coherence?

Integrating AI into the Operating Rhythm

- **Quarterly:** Analyze leadership communications and reviews for alignment across growth, cost, risk, and culture.
- **Annually:** Assess convergence of language and priorities across functions as an indicator of enterprise fluency.

- **Post-Major Change:** Evaluate whether crisis responses or transformation initiatives reinforce or fragment enterprise translation.

Over time, this feedback loop allows the organization to learn from its own patterns of language and decision-making. AI surfaces drift early, highlights misalignment before it hardens into culture, and strengthens judgment without centralizing control.

When embedded thoughtfully, AI helps translation become **self-correcting** and a property of the system rather than the burden of individual leaders.

Reflection

Would you be comfortable with AI replaying your words and decisions as a faithful representation of how leadership operates here? If not, what specific changes will you implement starting tomorrow to strengthen enterprise coherence?

12

CULTIVATING A LASTING LEGACY

W hen I reflect on my career from entrepreneurship in my early twenties, to corporate leadership roles, to my current work as a Chief Innovation Officer, a consistent pattern emerges. The moments that defined both my successes and my failures were never about programs, frameworks, or tools. They were about people.

They hinged on trust, on alignment with real business priorities, and on whether I was building capability beyond myself or unintentionally creating dependency, friction, or pressure that constrained others' ability to perform at their best.

That question sits at the heart of this book. This work is not about chasing a seat at the table. It is about becoming the kind of leader whose presence elevates the business because you have built successors, allies, and systems strong enough to sustain momentum without you. The contrast is easiest to see in practice.

Consider José Alvarez at Tucanes. His board presentations focused heavily on training hours delivered and completion rates achieved. When he requested additional headcount, the ask was denied. His influence stalled, not because he lacked competence or commitment, but because he framed his contribution in terms of activity rather than outcomes.

Now contrast that with Luis Gonzalez at Vida. From his first executive conversation, Gonzalez anchored his strategy to the CEO's priorities: attrition reduction, speed to performance, and responsible AI adoption. He didn't ask for a seat at the table. He earned one—by speaking the language of the enterprise. He was promoted to CLO with direct access to the CEO.

Two capable leaders. Two very different outcomes. The difference was not talent. It was translation, alignment, and business acumen.

The same pattern appears at scale. The Nuron rollout at Hilti was not treated as product training. It was positioned and executed as a transformation. Learning was scaffolded, advocates were intentionally cultivated, and the approach was grounded in the real challenges of the sales force. Adoption accelerated globally. Market share was preserved. Customer loyalty strengthened. And L&D's credibility shifted from service provider to strategic partner.

Or consider Leo McKnight. He was not the most senior leader at Hilti, but his influence shaped an entire generation of sales directors, general managers, and global leaders. His legacy was not a program he launched or a role he held. It was the people who carried his standards, judgment, and humanity into every corner of the organization.

These stories point to a simple, enduring truth: *Impactful leadership is defined by what you build and who can carry it forward.*

THE PATTERN ACROSS THIS BOOK

Across every chapter, a consistent pattern emerges. While the contexts differ, the underlying logic does not.

Business acumen is the foundation. Without a clear understanding of how the organization generates revenue, competes, manages cost, and grows, influence at the executive level remains limited. Fluency in the business is the entry point to credibility.

Talent development is the multiplier. Leaders who invest in others extend their impact far beyond their own role. They are remembered not for what they personally delivered, but for the leaders they developed and the capability they left behind.

Results are the language of power. Executives evaluate initiatives through the lens of growth, cost, risk, and enterprise value. When leaders fail to translate their work into these terms, relevance must be continually re-earned.

Allies are the force multipliers. Influence does not scale in isolation. Leaders who build strong alliances ensure their perspective is present even when they are not to amplify their impact across decisions and forums.

Governance is the guardrail. Good intentions and passion are insufficient. Disciplined decision rights, review mechanisms, and accountability structures protect strategy, reduce friction, and reinforce trust.

Culture is the accelerant. Strategy succeeds or fails based on how people behave. Leaders do not mandate culture; they model it, reinforce it, and reward it through consistent signals.

Technology is the ecosystem. Tools such as AI or TMS platforms are not solutions on their own. They become multipliers only when integrated with people, culture, and strategy amplifying judgment rather than replacing it.

Legacy is the measure. Leadership is never neutral. What you tolerate, what you reward, and who you develop determine what endures long after your role changes.

That is the arc of this book. From reframing training as transformation to cultivating leaders who can carry that transformation forward, the message is consistent: **credibility is built by connecting work to the business, and legacy is built by developing people who can sustain that connection.**

Zooming out further, each chapter contributes to a clear human continuum:

- Business acumen helps individuals understand how their work fuels the enterprise.

- Talent development builds leaders capable of consistent judgment, not just task execution.
- Allies and governance create shared accountability and enterprise alignment.
- Culture and technology amplify sound decision-making rather than compensate for its absence.
- These principles become embedded in people allowing them to be recreated in any environment.

Ultimately, the most enduring outcome is not the framework itself, but the leaders who carry the capability forward. Systems may change. Strategies will evolve. What lasts is the ability to think, decide, and lead effectively wherever the next challenge appears

The Consequences of Neglecting This Approach

What happens when leaders fail to operate this way? We do not need to speculate. The evidence is already visible.

Marginalization of Functions

RedThread Research reports a 50% decline in L&D's involvement in strategic conversations over the past two years. When leaders fail to connect their work to measurable business outcomes, they are gradually excluded from the decisions that shape the future of the enterprise.

Erosion of Influence

McKinsey research shows that organizations filling fewer than 30% of senior roles internally are twice as likely to underperform financially. Weak leadership pipelines do not merely slow progress; they increase organizational fragility during moments of pressure.

Fractured Culture

Deloitte's 2025 research identifies cultural misalignment as the leading cause of failed transformations. Strategy alone does not fail— execution fails when leaders neglect the cultural conditions required to sustain change.

These are not abstract risks. In volatile environments, executives fund what delivers results and protects resilience. Functions and leaders that cannot demonstrate enterprise value are deprioritized regardless of effort, intent, or historical contribution.

Every executive therefore faces a clear choice: optimize for short-term activity or build long-term capability. Every decision, conversation, and tolerance signal contributes to that outcome. Excusing weak leadership, prioritizing activity over outcomes, or avoiding account-ability may preserve comfort in the moment but it cultivates fragility. Modeling courage, aligning work to business results, and developing others builds organizations that endure.

Legacy is not about resisting change. It is about constructing foundations strong enough to absorb it.

A CALL TO ACTION

Over the course of my career as a head of learning and later as a Chief Innovation Officer I learned a difficult truth.

> **Learning is only valuable when it drives business outcomes. Innovation is only meaningful when it is disciplined, governed, and culturally aligned.**

Leadership is not defined by what you do, but by what you build through your decisions and the behaviors that you consistently reinforce.

The true measure of leadership is visible in the ability to navigate ambiguity, act with ownership, balance performance with empathy, and communicate credibly in the language of results long after your direct involvement ends. That is the legacy that matters.

- Build systems where technology amplifies judgment, rather than replacing it.
- Develop leaders who think beyond functional silos and act for the enterprise.
- Cultivate allies who extend your influence when you are not in the room.
- Establish governance that channels passion into sustained execution.
- Foster cultures where care and performance reinforce rather than compete with each other.

Your impact will be measured by the leaders you leave behind, by the culture they sustain, and by the enterprise that continues to perform because you built capability, not dependency.

Choose to build. Always.

13

APPENDIX: SELF-ASSESSMENT
LEADING WITH BUSINESS OUTCOMES

This self-assessment is designed for executives and senior leaders seeking to evaluate the degree to which their functional area is aligned with overarching business outcomes. This is not an evaluative assessment, but rather a reflective tool to identify strengths and areas for development.

Instructions: Rate each statement on a scale of 1 to 5.

- 5 = Consistently true
- 4 = Often true
- 3 = Sometimes true
- 2 = Rarely true
- 1 = Not at all true

SECTION 1: GROWTH

I directly link my team's work to revenue growth, market expansion, or customer acquisition.

I measure improvements in key performance indicators such as win-rate, conversion, adoption, or customer lifetime value, rather than solely focusing on activity levels.

I can clearly articulate how our initiatives accelerate time-to-market or shorten ramp-up time for new employees.

SECTION 2: COST

I can quantify the cost savings resulting from my team's initiatives (e.g., reduced turnover, rework, or inefficiencies).

My reporting emphasizes operating margin and contribution to profitability, rather than solely focusing on activity metrics.

I reinvest savings into higher-value initiatives that are strategically aligned with organizational goals.

SECTION 3: RISK

I proactively anticipate risks that could impede strategy execution (regulatory, reputational, operational, or talent-related).

I can demonstrate how our work mitigates or prevents these risks.

I utilize data-driven insights, rather than anecdotal evidence, to demonstrate risk avoidance or reduction.

SECTION 4: SHAREHOLDER VALUE

I connect our work to enterprise value, succession stability, or investor confidence.

I can demonstrate how people-related initiatives protect or enhance long-term brand equity.

I can explain how my function maintains confidence in our business model during periods of disruption.

SECTION 5: BUSINESS ACUMEN

I read and interpret financial statements (P&L, balance sheet, cash flow) in the context of my team's work.

I can explain how changes in market share, pricing, or supply chain costs impact my area of responsibility.

My team is comfortable discussing revenue, margin, and growth in clear and accessible terms.

SECTION 6: TALENT DEVELOPMENT

I deliberately provide stretch assignments, coaching, and development opportunities for potential successors within my team.

I rotate individuals across projects or functions to broaden their perspectives and skill sets.

I consider leadership development an integral part of my role, not a discretionary activity.

SECTION 7: ALLIES & INFLUENCE

I regularly engage with peers in Sales, Finance, Operations, Product, and HR to ensure alignment on shared priorities.

I proactively cultivate relationships before they are needed.

I have established relationships with individuals who would advocate for my initiatives in executive meetings, even in my absence.

SECTION 8: CULTURE & GOVERNANCE

I understand how our company's culture enables or impedes change initiatives.

I model and reward behaviors that reinforce a strong organizational culture.

I actively participate in governance processes to ensure initiatives align with strategic objectives.

SECTION 9: TECHNOLOGY & AI AS A MULTIPLIER

I consider technology as an integral part of the business ecosystem, rather than a standalone solution.

I leverage AI and analytics to identify correlations between initiatives and outcomes.

I utilize pilot programs with leading indicators and well-defined exit criteria to evaluate new technologies.

SCORING & REFLECTION

- 100–135 points (avg. 4–5): You consistently demonstrate business-outcome-driven leadership. Your challenge lies in scaling this approach by empowering others to think and act similarly.
- 70–99 points (avg. 3–4): You demonstrate a good understanding of the connection between activities and outcomes, but with areas for improvement. Develop an action plan focused on the 2–3 sections where your score was lowest.
- Below 70 points (avg. <3): Your leadership communication may be primarily activity-focused. Prioritize developing a strong understanding of growth, cost management, risk mitigation, and shareholder value before attempting to influence strategically.

Next Step

Identify your lowest-scoring section and draft one specific action you will take within the next 30 days to enhance your leadership in that area.

www.ingramcontent.com/pod-product-compliance
Lightning Source LLC
Chambersburg PA
CBHW072307210326
41519CB00057B/3043